Exploring the History of Nuclear Physics

AIP Conference Proceedings

Series Editor: Hugh C. Wolfe

Number 7

CONFERENCES

Exploring the History of Nuclear Physics

(Proceedings of the American Institute of Physics - American Academy of Arts and Sciences Conferences on the History of Nuclear Physics, 1967 and 1969)

Edited by

Charles Weiner, assisted by Elspeth Hart

Center for History of Physics
American Institute of Physics

American Institute of Physics

New York 1972

Library of Congress Catalog Card

Number 72-81883

American Institute of Physics

335 East 45th Street

New York, N. Y. 10017

Printed in the United States of America

CONTENTS

Foreword

Scientists tend to be uninterested in history, even the history they have made, and are usually content to let the scientific record speak for itself. Such an attitude probably comes from the feeling that what is permanent or true or useful about the life-work of all scientists exists as a part of the state of scientific knowledge at any time, whereas the historical record displays errors or inaccuracies or outmoded or archaic theories, concepts, or experiments that either led to blind alleys or just didn't turn out to be useful. Whatever the reason, however, the fact remains that scientists, unlike diplomats or statesmen, are generally so uninterested in history that all too often they do not even bother to preserve the memoranda, correspondence, notebooks or laboratory journals which document the development of their own science. Only occasionally, most often when questions of priority of discovery or invention arise, do scientists evince any concern for history, although there are some notable exceptions of scientific autobiographies or of perceptive biographies of one scientist written by another.

It is even more astonishing that historians of science have so concentrated on the science of the past that they have neglected to examine at first-hand the formation of science in our own time. I am continually struck by the fact that historians did not consider it a primary obligation collectively to seek out first-hand information from the founders of the scientific revolution of the twentieth century: interviewing the founders of atomic and nuclear physics, relativity theory, and quantum theory, and collecting their correspondence and papers. So far as I know, no one competent in science and in history ever made a complete study of a major figure in twentieth century science prior to 1945, by systemically exploring with him over a long interval the real stages of the development of his thought. Only in recent years, and much too late, have programs finally been mounted to obtain such first-hand information, but too late for extended interviews with such figures as Albert Einstein, Henri Poincaré, E.O. Lawrence, Wolfgang Pauli, Lord Rutherford, or J. J. Thomson, to name but a few individuals of the many who have forged the revolution in the twentieth century.

The first instance which I ever encountered of the failure to obtain such information was on reading a biography of Newton written by William Stukeley, in 1752, but not published until 1936. Stukeley, was, as he said, "a countryman of Sir Isaac Newton's," and knew Newton personally. Stukeley not only learned the facts of Newton's life from records and interviews with Newton, but he also talked with a number of men who had been associated with Newton, including the amanuensis who wrote out the text of his famous *Principia.* Many interesting facts about Newton were recorded, such as, "I have heard him say that, during his closest application, he never forgot going to bed about 12. This he learn'd by experience; for if he exceeded that hour, it did him more mischief in his health, the next day, than a whole day's study at regular times."[1] What is most apt to attract the attention of a historian reading this biography is the account of Newton's most famous discovery, the principle of universal gravitation. Stukeley records that he and Newton, after dinner, were sitting in the garden drinking tea, "under the shade of some appletrees, only he and myself." Then,

. . . Amidst other discourse, he told me, he was just in the same situation, as when formerly, the notion of gravitation came into his mind. It was occasion'd by the fall of an apple, as he sat in a contemplative mood. Why should that apple always descend perpendicularly to the ground, thought he to himself. Why should it not go sideways or upwards, but constantly to the earths centre? Assuredly, the reason is, that the earth draws it. There must be a drawing power in matter: and the sum of the drawing power in the matter of the earth must be in the earths center, not in any side of the earth. Therefore dos this apple fall perpendicularly, or towards the center. If matter thus draws matter, it must be in proportion of its quantity. Therefore the apple draws the earth, as well as the earth draws the apple. That there is a power, like that we here call gravity, which extends its self thro' the universe.

1. William Stukeley, *Memoirs of Sir Isaac Newton's Life: Being Some Account of His Family and Chiefly of the Junior Part of His Life, 1752.* Edited by A. Hastings White. (London: Taylor and Francis, 1936), p. 61.

And thus by degrees he began to apply this property of gravitation to the motion of the earth and of the heavenly bodys, to consider their distances, their magnitudes and their periodical revolutions; to find out, that this property conjointly with a progressive motion impressed on them at the beginning, perfectly solv'd their circular courses; kept the planets from falling upon one another, or dropping all together into one center; and thus he unfolded the Universe. This was the birth of those amazing discoverys, whereby he built philosophy on a solid foundation, to the astonishment of all Europe.[2]

When I first read that statement, and when I have reread it as I have many times since then, I have thought to myself that here was one of the greatest opportunities to find out in depth and in full detail how one of the greatest men of science had conceived that he had made his discoveries. We, with only such statements as this to guide us, must use our imaginations to reconstruct the pattern of Newton's thought but Stukeley was in a position to have pushed the enquiry further, had he only had the wit and the historical sense to recognize the uniqueness of the opportunity afforded him, and to have given us Newton's answer to questions about Newton's concepts that are today of such concern to historians and philosophers and scientists. Stukeley may provide a cautionary example to historians to be on their guard lest their posterity say that they will have missed a golden opportunity.

In recent years, some historians and scientists, notably physicists, have been actively seeking to preserve the records of the science of our time, and — however belatedly — making taped interviews which we hope may be better sources of information for the future than the account of Stukeley's interviews with Newton. A single interview, however, is a very limited historical technique, and is apt to be only a device whereby a scientist presents a re-ordering of his own experience, unconsciously conforming (in all too many cases) to some preconceived idea as to how science in general, and his science in particular, should be produced. Thus it is of extreme value to bring together historians and scientists to explore fundamental developments

2. *Ibid.*, pp. 19-20.

of science, such as those of nuclear physics, for the historians may then push whole groups of scientists into recalling their collective memory, and the interactions of one scientist with another may elicit the several different dimensions of historical truth. The text before us is a first essay into such historical recall. All of us who participated in it trust that similar conferences may be held in other disciplines, and in other aspects of physics, leading us step by step to a better understanding of the nature of science, the process of discovery and the dimensions of the scientific revolution of our time.

I. Bernard Cohen
Professor of History of Science
Harvard University

When I agreed to participate in the AIP-American Academy project on the history of nuclear physics I did so on the basis of both my professional responsibility and my personal curiosity. Nuclear physics, like other dynamic fields of physics, is an important part of the social and cultural history of our time. Physicists have a responsibility to aid historians and sociologists in documenting and studying this history and the project was conceived as an interesting and pleasant experiment to explore how we could provide such assistance. Besides, those of use who have been active in physics during the past four decades rarely have an opportunity to talk and reflect about the events and experiences of that period. It is always pleasant to talk about physics, and especially to talk about physics in the old days. When such a pleasant activity is also useful, so much the better. Scientists are curious about the unknown, and the project provided a pleasant and useful way of satisfying our curiosity by increasing our own understanding of the relatively unexplored history of nuclear physics.

The interviews and the conferences stimulated memories of colleagues and of the excitement and atmosphere of earlier years. We learned a lot from the recollections of one another and from the discrepancies that came to light. Cross-examination by the historians, especially when they offered letters and other documents as evidence, brought out even more information. The cooperative nature of the enterprise and the new sources of information it has unearthed have made it a most worthwhile experience.

As readers of these proceedings can readily see, the conference discussions do not constitute a history of the field of nuclear physics. They lack balance, contain many gaps, and reflect the experiences and biases of the participants. A different group would have provided a different picture. However, the proceedings record our mutual efforts to probe into ways to properly document the complex development of this branch of physics. The discussions were only the initial exploratory steps in what must be a continuing inquiry. For example, one of the conferences was focused on the role of theory, not because of any implications that theory had played the leading role in the development of nuclear physics, but because the historians felt that it would be useful to explore how the subject looked to the theorist in the 1930s and 1940s. Similar explorations of the central role played by the experimentalists in bringing nuclear physics forward are essential for any real understanding of what happened.

The ultimate value of these conferences will be in their stimulation of further inquiry into the history of nuclear physics and of other fields of physics, as well as of fields in other scientific disciplines. It is gratifying to know that such inquiries are underway, because the scientific community and the general public have a greater need than ever to understand the processes of recent scientific change.

Victor F. Weisskopf
Professor of Physics
Massachusetts Institute of
Technology

Introduction

The following pages are the record of part of a unique experiment in the historiography of twentieth-century science. They contain transcripts of the discussions at two small conferences in which the participants probed into the relatively unexplored recent history of a field of physics to identify needs and opportunities for documentation and study of its development. The participants included physicists who had played roles in the development of nuclear physics during the 1930s and 1940s, the decades which encompassed its rapid growth, maturation, and initial applications. Joining the physicists in these discussions were historians, philosophers and sociologists who were either undertaking studies of the history of nuclear physics or were involved in research on related aspects of the development of twentieth-century science. Thus, the historical actors in a recently spawned and still evolving field of knowledge came together with the scholars concerned with documenting, studying, and analyzing it. Together they attempted to identify questions of historical interest, offered recollections or interpretations of specific events and processes, and suggested sources and methods of historical documentation so that serious study of these developments would be possible in the future.

The lively discussions provided valuable historical information, insight, and interpretation. They have paved the way for several historical studies now underway and have led to the location and preservation of original source materials not previously available for research. Aside from these substantial results, the conference discussions have an intrinsic interest because they reveal something of the shared values of the physicists, and the fascinating interplay of individual recollections within this collective effort to reconstruct the circumstances and dynamics of past events. They are presented here, with some supplemental documentation and appendices, as the record of early stages in an experiment that is still in progress.

The transcripts of the discussions can be more fully understood and evaluated when the conferences are viewed in the context of the project of which they were a part. The conferences were organized in connection with a joint American Institute of Physics - American Academy of Arts and Sciences project to explore the needs and opportunities for historical research on fields of physics which have emerged and developed rapidly during the past four decades and which have had substantial social impact within the physics community and in the larger social communities in which they function. The project was initiated in 1966 when the American Institute of Physics and the American Academy of Arts and Sciences established a joint committee on the history of contemporary physics consisting of Oscar Handlin, Gerald Holton, Robert Merton, Philip Morrison, Cyril Smith and Victor Weisskopf, with I. Bernard Cohen as chairman and myself as project director. Since 1967 Victor Weisskopf and I. Bernard Cohen have been co-chairmen. The composition of the committee—historians, physicists, and a sociologist—reflects the interdisciplinary nature of the enterprise.

Recognition of the need for such a project, and confidence that it could be undertaken, stemmed from the experiences and results of the AIP program in the history of physics which, since 1961, had been actively engaged in efforts to locate, preserve and encourage the use of source materials to document the development of twentieth century physics. By 1966 the AIP Center for History of Physics had been successful in finding a wealth of original historical documents and had arranged for the deposit and cataloging of these letters, notebooks, and diaries at a number of university and private archives throughout the United States. A series of oral history interviews with physicists was underway, a National Catalog of Sources for History of Physics had been established and was rapidly growing, and the research materials and information resources of the Center's Niels Bohr Library in the AIP headquarters building in New York City were being increasingly used by scholars and educators. All of this had been achieved through the enthusiastic cooperation and collaboration of physicists, historians, sociologists, philosophers and archivists at scores of institutions in the United States and in Europe.

These results, resources and relationships pointed to the desirability of initiating an exploration of the historical development of the major fields of physics research which have grown so rapidly and with such great social impact since the early 1930s. What were the interesting historical questions? What lines of inquiry promised to be fruitful in leading to an understanding of the social and intellectual processes involved in this aspect of recent scientific change? What documentary sources would be needed as the basis for such studies, and where could they be found? Perhaps it would be possible to suggest answers to these questions by focusing on the emergence and growth of a specific field. Such exploration might stimulate needed historical and sociological research and would hopefully lead to the identification and preservation of source materials before they were scattered, destroyed or lost. Whatever was learned from the project could be applied to help develop guidelines for the larger continuing effort to document and study the development of all of twentieth century physics.

With these aims in mind the joint AIP - American Academy project was launched. The field of nuclear physics was selected for the initial explorations because it had become clearly defined as a major research field in the 1930s, had developed especially rapidly in the United States, and had far-reaching social applications during and since World War II. In addition, nuclear physics contributed to the birth of particle physics, another field which brought about major changes in the scale, organization, style and cost of physics research. Many of the individuals who had played important roles in these events were still active in physics and a large number of them resided in the United States. Nuclear physics appeared to offer fruitful possibilities for the project's first efforts.

Staff work on the project began at the Center for History of Physics to prepare for an initial conference of physicists who had been involved in the development of nuclear physics and historians, philosophers and sociologists currently involved in research on 20th-century science. Published and unpublished source materials were studied and tape-recorded interviews were conducted with an initial group of 13 physicists selected because of their own contributions and/or knowledge of the development of the field at various institutions. The first conference was held on May 18 and 19, 1967 in the congenial environment of the American Academy of Arts and Sciences in Brookline, Massachusetts. Half of the twenty-five conference participants were nuclear physicists and half were either historians, philosophers, or sociologists of science. Five of the physicists were among those who had been interviewed. A discussion paper (See Appendix I), distributed prior to the conference, reported on the preliminary exploration of factors involved in the initiation and growth of nuclear physics, based on information and insights obtained through the survey of documents and from the interviews. The report was intended to stimulate discussion by raising questions which appeared to be of historical and sociological interest, identifying matters in need of clarification, and outlining possibilities for documentation and study.

The free-wheeling discussions at the conference were organized around three general themes: the initiation of nuclear physics as a major field of research; the interaction of theory and experiment; and the postwar development of the field. The conference brought a considerable amount of information to light. It also demonstrated that historical and sociological research on nuclear physics was both needed and feasible and that the time was ripe for identifying and preserving relevant source materials, gathering quantitative data, and conducting specific historical studies to shed light on developments of major significance. The participants also recommended a continuation of the program of tape-recorded interviews with individuals whose recollections could provide valuable information and insight and could lead to the preservation of documentary materials.

These objectives were furthered late in 1968 when the National Science Foundation approved the committee's application for a grant in support of research under my direction at the AIP Center for History of Physics. Collaborating in this research from the beginning of 1969 was another historian of science, Henry Small, and we were joined in September 1969 by a sociologist, Beverly Porter. The aim of the research studies was to provide data on the overall growth of nuclear physics as a field, to identify stages in its development, and to trace and analyze the conceptual, technical, and social aspects that were involved.

In connection with these research and archival activities, the joint committee sponsored a second conference at the American Academy on May 18 and 19, 1969. The conference focused on the role of theory and theorists in the development of the field from about 1930 to about 1950. It provided a unique opportunity for some of the physicists who participated in these developments to discuss, on the basis of their own experience, the theoretical ideas which influenced the overall growth of the field. We wanted to explore how and why concepts were introduced, how they were received at the time, and the role they played in subsequent research. As background material for the conference, participants were sent a compilation of relevant extracts from the transcripts of some of the interviews and from letters written during the period under discussion. The extracts were selected and arranged to give glimpses of the motivations, circumstances and impact of the introduction of various concepts. Some of the accounts were juxtaposed; some were arranged to illustrate different stages in the development and use of the concepts. Several of the physicists at the conference were among those who had been interviewed and the discussions provided them with a unique opportunity to compare their partial perspectives with those of their colleagues. The discussions were as lively and informative as those at the initial conference and provided additional insights and leads for documentation and research.

In the course of the subsequent research original source materials, documenting the work of individuals and the development of the field in specific institutional settings, were located and studied; data on the physics literature and on doctoral dissertations in the field was gathered and analyzed; and additional interviews were conducted with physicists in the United States and in Europe. Results of the research have been presented in several work-in-progress reports and articles.* More detailed and analytic studies are now being completed. The appendices to these conference proceedings outline the resources for further research that have been located, compiled, or created by the project staff. The data, catalogs, interview transcripts, and documentary sources will be preserved and made available for other researchers in the future as part of the resources of the AIP Center for History of Physics.

It is clear that the conferences have already had a substantial "multiplier effect" by stimulating historical and sociological research and by aiding in the identification and preservation of original source materials. But what can be learned from the proceedings *per se*? Surely they do not constitute a history of nuclear physics. Since no formal papers were presented at the conferences, the mode of communication was exclusively through open discussions. The proceedings include the factual mistakes, historical gaps, and lack of balance inevitable in such discourse. A different group of participants would have provided different perspectives based on their own experiences and interpretations. But their contributions to the discussions would not be fully accurate or complete either. In the spirit of the experiment, neither the participants nor the editors have tried to "clean up"

---

*Charles Weiner, "A New Site for the Seminar: The Refugees and American Physics in the 1930's," in B. Bailyn and D. Fleming, eds., *Intellectual Migration* (Harvard University Press, 1969), pp. 190-234; originally appeared in *Perspectives in American History, II,* 1968.

__________, "Prospects for Research on the History of 20th Century Physics," *Proceedings of XII International Congress of the History of Science, 1968,* Tome V, (Blanchard, 1971), 111-115.

__________, "Physics in the Great Depression," *Physics Today* (October 1970), 31-37.

__________, "1932 – Moving into the New Physics," *Physics Today* (May 1972), 40-49.

__________, "Institutional Settings for Scientific Change: Episodes from the History of Nuclear Physics," in A. Thackray and E. Mendelsohn, eds., *Science and Values* (Humanities Press), in press.

Henry G. Small, "Nuclear Physics in *The Physical Review,*" work-in-progress report, History of Science Society, December 1970.

__________, "The Rise of Nuclear Physics in the United States," work-in-progress report, History of Science Society, December 1971.

Beverly Porter, "The Seeds of Big Science: U.S. Nuclear Physics 1927-1940," paper to be presented at International Sociological Association, London, September 1972.

Conference participants who have since become concerned with the history of nuclear physics include Joan Bromberg (see her article, " The Impact of the Neutron: Bohr and Heisenberg," *Historical Studies in the Physical Sciences, 3,* 1971, 307-341), and Barbara Reeves Buck who is engaged in a doctoral dissertation study of the Rome group of nuclear physicists in the 1930s.

the transcripts of the discussions. We felt that the interactions at the conferences helped to increase the historical understanding of all of the participants and we wanted to share this experience with others.

Our discussions explored the development and role of individuals, ideas, and institutions, emphasizing that the history of science, like that of other human endeavors, is richly complex. We noted that physicists who specialized in experimental work often had motivations and perceptions different from those who had specialized in the design and development of scientific instruments or in theoretical studies. We learned that an awareness of these differences can help to increase understanding of the processes of scientific creativity and change. Such understanding is often obscured if we limit our inquiries to "physicists," or even to "nuclear physicists." We also discussed the meaning and validity of the physicists' retrospective statements that it had been "obviously clear" to all knowledgeable individuals what the most important problems were and which ones were ripe for attack. Many other themes are evident in these proceedings, lending added interest to the rich, personal accounts that provide its main substance.

Now some words about the organization of this volume. The proceedings of the May 1967 conference constitute Part I: Emergence and Growth of Nuclear Physics as a Research Field. The proceedings of the May 1969 conference constitute Part II: The Role of Theory in the Development of Nuclear Physics. Each Part is subdivided into Sessions corresponding to the general, but not exclusive, focus of the discussion. All of the discussions were tape-recorded and transcribed verbatim. The original tapes and transcripts are preserved at the AIP Center for History of Physics. A minimum amount of editing was done, only enough to eliminate confusion in the reading of the transcript. Some portions of the discussion dealing with the organization of the conferences and such matters as the techniques of interviewing have not been included in these proceedings. The transcripts were sent to the participants who were asked, in the spirit of the experiment, to limit their corrections to obvious transcribing errors. Then citations were added at appropriate points to provide information on specific papers under discussion. Interspersed in the proceedings are Heka Davis' candid photographs of the participants, providing glimpses of the atmosphere of the meetings. A small section of historical photographs, between Parts I and II, provides some of the flavor of the earlier interactions that were the subject of our discussions.

Appendix I is the discussion paper prepared and distributed prior to the first conference. It was effective in provoking discussion and is reproduced here as part of the record of that meeting, even though its authors would now approach these questions differently on the basis of their subsequent research. Appendix II, prepared by Beverly Porter and Henry Small, describes some of the data resources assembled as part of the research project stimulated by the conferences. Appendix III lists some of the collections of original source materials located and studied in the course of the project and describes the oral history interviews, including their nature, scope and potential use for historical research.

Together, the proceedings and appendices record the initial stages of an effort to explore the history of recent science. This effort has been made possible by a number of individuals and institutions who either were part of the experiment, helped to carry it out, or provided support for it. As an historian who has enjoyed the human experiences of the project and believes he has increased his understanding of the history of science from it, I thank them all. Essential to the entire project was the wholehearted cooperation of the many individuals who agreed to participate in the interviews and of those who attended the conferences and contributed to the discussions. The initial conference was supported by a grant from the Committees on Research Funds of the American Academy of Arts and Sciences, and by the American Institute of Physics. Research grants from the history and philosophy of science section of the National Science Foundation provided support for the subsequent research project of which the second conference was a part. Members of the joint AIP - American Academy Committee on the History of Contemporary Physics consistently offered imaginative and thoughtful guidance in the organization and carrying through of the conferences. The congenial atmosphere, superb facilities and helpful staff of the American Academy of Arts and Sciences provided an ideal environment for the conferences.

Henry Small played an important role in the research preparation for the second conference and in the related research project at the AIP Center for History of Physics. The project has also benefited in a major way from the sociological studies of Beverly Porter. Special studies for the preparation for the first conference and its follow-through were carried out by Lewis Auerbach, Joan Bromberg, Elizabeth Garber, Barry Richman and John Titus. Research assistance for the project has been provided by Dorothy Schreibersdorf and Elspeth Hart. The source materials and information resources of the Center for History of Physics' Niels Bohr Library have been essential for the project, and have been made available through the special efforts of Joan Warnow, with the assistance of Elizabeth Barton.

The preparation of these proceedings and the publication of this volume would not have been possible without the dedicated efforts of Elspeth Hart who skillfully and thoughtfully saw it through each stage from raw transcript to completed manuscript.

Charles Weiner, Director
Center for History of Physics
American Institute of Physics

Proceedings of the

American Institute of Physics - American Academy of Arts and Sciences

First Exploratory Conference on the History of Nuclear Physics

May 18, 19, 1967, Brookline, Mass.

EMERGENCE AND GROWTH OF NUCLEAR PHYSICS AS A RESEARCH FIELD

Participants at May 1967 Conference. (Photos by Heka.)

PARTICIPANTS **May 1967 Conference**

Lewis Auerbach, Department of History of Science, Harvard University
Garrett Birkhoff, Department of Mathematics, Harvard University
Joan Bromberg, Department of Social Sciences, Polytechnic Institute of Brooklyn
Peter A. Caws, Department of Philosophy, Hunter College of the CUNY
I. Bernard Cohen, Department of History of Science, Harvard University
Robert S. Cohen, Department of Physics, Boston University
Oscar Handlin, Charles Warren Center for Studies in American History, Harvard University
Raymond G. Herb, Department of Physics, University of Wisconsin
Erwin N. Hiebert, Department of History of Science, University of Wisconsin
Gerald Holton, Department of Physics, Harvard University
Daniel J. Kevles, Division of Humanities, California Institute of Technology
H. William Koch, American Institute of Physics
Thomas S. Kuhn, Program in the History and Philosophy of Science, Princeton University
M. Stanley Livingston, Department of Physics, Massachusetts Institute of Technology
Everett Mendelsohn, Department of History of Science, Harvard University
Robert K. Merton, Department of Sociology, Columbia University
Philip A. Morrison, Department of Physics, Massachusetts Institute of Technology
Barbara Reeves, Department of History of Science, Harvard University
Barry Richman, Center for History of Physics, American Institute of Physics
Emilio G. Segrè, Department of Physics, University of California, Berkeley
Dudley Shapere, Department of Philosophy, University of Chicago
Cyril S. Smith, Departments of Metallurgy and Humanities, Massachusetts Institute of Technology
Valentine L. Telegdi, Department of Physics, University of Chicago
Merle A. Tuve, Department of Terrestrial Magnetism, Carnegie Institution of Washington
John Voss, American Academy of Arts and Sciences
Charles Weiner, Center for History of Physics, American Institute of Physics
Victor F. Weisskopf, Department of Physics, Massachusetts Institute of Technology
John A. Wheeler, Department of Physics, Princeton University
Robert R. Wilson, Department of Physics, Cornell University
Harriet Zuckerman, Bureau of Applied Social Research, Columbia University

Affiliations shown are those at the time of the 1967 Conference

Session I. The Initiation of Nuclear Physics as a Major Research Field

Cyril S. Smith, Chairman

Smith: Nuclear physics has not been a popular subject for historical attention for very long. Sometimes the thought passes through my mind that the concern with history, except for professionals, is usually the mark of an old man. Is it possible that physics has reached the post-mature stage? I hope not.

However, the purpose of this meeting is much more to raise questions and see what the important ones are rather than to settle anything. It seems to me that there are three aspects of what we are after here. Foremost, of course, the facts of what happened are the things on which one has to base any speculation, and one must have a close concern with these. On the other hand, facts by themselves are useless.

We also should consider at this meeting the question of what are the records we want to keep. What are the records from the past that we want to look at? I was much impressed by Emilio Segrè's remarks last night when he came to realize that a record that he himself had made [a tape-recorded historical interview made in 1967] was not what he thought it should have been. I wonder whether he would also have thought a record made 20 years ago was a correct record. The uncertainty of records and realizing the things which are left out, is so very basic to our understanding history.

Then there is another point. It is well known, of course, what the difference is between a discipline like history and a discipline like physics. The physicist is able to isolate the things that he is going to talk about. He does not say other things are not important, but he says, "For the time being I'm going to look at one thing." He can be simple. He can use fairly definite methods for checking whether his ideas are right or not. History essentially is complex, and if you take away the complexity, you have no history. The two approaches must be very different.

Proceedings of the Conference on May 19, 1967. On the previous evening there was an informal dinner at the American Academy followed by introductory remarks about the purposes of the Conference. During the evening a recording of a speech by Rutherford in 1931 was played and Emilio Segrè showed slides of candid photographs of physicists at work and at meetings in the 1920s and 1930s.

Incidentally, this business of complexity and history is completely interwoven. There cannot be even a complex piece of matter which has its structure and its character except as a result of its history. Possibly solid-state physics, when it gets a little more complex than at the moment, will be a subject closer to the historian's taste than nuclear physics.

We are concerned at this meeting with the history of nuclear physics. We must keep straight, I think, the difference between the nucleation of nuclear physics and the growth of nuclear physics. Again, to take an analogy from solid-state physics: the beginning of something and its growth call for quite different considerations, just as men who start revolutions are usually not the ones to run the government afterwards. There are many wonderful interacting levels of intellectual and sensual and social factors that are involved in any human activity. I hope the relative importance of these things will come out during the discussion.

It is essential, I think, at this meeting to realize that the scientists are the ones who have to provide a good deal of the data and of the viewpoints which the historians and the sociologists will talk about. The broad data will be mostly from the scientists; yet the historians and the sociologists have a very important role in showing that the things that happened are probably not as simple as the physicists think they are, particularly in retrospect. So the historians and the sociologists must be very free to break in and to point out the fallacies in the simplifications which the physicists will inevitably use. And, equally true, the physicists must remember that the historians in trying to talk history are themselves over-simplifying to an extreme extent. And the historians' simplified patterns are open to criticism just as much as the physicists'.

The center of the first section of our discussion this morning is Professor Weisskopf's introduction.

Weisskopf: First of all, it is not the center; it is a beginning. I emphasize this because I have not and could not prepare a systematic thing. Second, I am glad you mentioned the fact that we physicists are so unhistorical. In this respect I am an extreme physicist, and also I have a very bad memory, which is connected with my unhistorical attitude. Once, in high school, the history teacher was dissatisfied with me and said, "You don't know any dates." And I said, "I know all dates. I only don't know what happened on them."

I also would like to say something else before I really start. I had thought, and this is not a reproach to Charles Weiner, that this meeting would be different. You see, there are several of us who were interviewed. I thought that we were going to read each other's interviews and then come together, watched by a few of you historians and sociologists, and we would find out how differently we looked at it. Then we would try to straighten out the opposing statements that we found in those interviews, as I am sure there must be many. Now, that is probably one of the next steps and might be a very useful one for the experience and also for the use of the historian.

At present we are not yet that far. I, at least, as one of the interviewees, have not seen any other interview. It may be true for all. I know that my interview, for example, was very subjective. As long as I have no feedback I'm even more subjective. That is one of the troubles.

Now, I would like to make a very few informal remarks, which are mostly questions. So when I make a statement, it is meant really as a question.

Smith: That goes for everybody at this conference, I hope.

Weisskopf: I hope there will be some statements: for example, at what time what paper appeared and what did it say—things of this kind for me are still questions.

Well, let me now start in a more or less unsystematic way. I will use the study paper's two divisions for my section—namely, internal history, and the social aspects of it, the exterior situation.* As far as the former is concerned, it is generally said that nuclear physics was born or really took a strong momentum in 1932, the great year when the neutron was discovered, accelerators were built, the deuteron was discovered, and artificial radioactivity was about to be discovered. This is only true—again, this is a question—in a restricted sense. There was certainly nuclear physics before. Yesterday we heard a little tape, and if that is not nuclear physics, I do not know what is. Rutherford

*See Appendix I, p. *250*.

was talking [in 1931] about the gamma-ray spectra of the nuclei and was trying to investigate them, and that is certainly nuclear physics.

May I remind you of something I think is very important? Otto Stern measured the magnetic moment of the proton, which is the beginning of a very important branch of nuclear physics. And here is a real question to the expert: I think measurements of magnetic moments of nuclei by any methods, hyperfine structure and so on, must have been done before 1932.

Physicists: Sure, sure.

Segrè: I was in Hamburg at the time.

Weisskopf: You measured it?

Segrè: No, I didn't measure it, but I was in Hamburg when Stern was doing it, and of course, Fermi and I wrote papers on hyperfine structure.

Weisskopf: That is right and there was the Fermi theory of hyperfine structure.

Telegdi: Excuse me, Viki. Nuclear spin was discovered before the electron spin—in 1924. Pauli's paper is 1924.[1] In that paper he specified that the nucleus has a moment in order to interact and produce fine structure.

Weisskopf: Now, this is a very important point in nuclear physics. But of course something happened in 1932. I suppose that what one has in mind essentially is that the discovery of the neutron and the deuteron gave a tremendous push for the better understanding of the nucleus. And before 1932 one was starting in the dark.

1. Pauli, "Zur Frage der theoretischen Deutung der Satelliten einiger Spektrallien und ihrer Beeinflussung durch magnetische Felder," *Naturwissenschaften, 12,* 741, (August 1924).

Now, again, I would say that the question for this gathering is: How far is this really true? For example, how many of some of the bright ideas were not already expressed before? I would like to tell one story, which seems to me characteristic. It happened I think in the spring of '32, a few months before the discovery of the neutron. I was with Heisenberg in Leipzig, with Bloch and Placzek and a lot of other people there. And we were sitting in a coffee house which happened to be very near an indoor swimming pool building. And we were talking about beta decay and the question: Where is the electron? What is the electron doing in the nucleus? And Heisenberg pointed at the swimming pool building and said, "Look, here people go in and come out all dressed. Do you conclude from this that they also swim dressed?" So, evidently, for this problem of "What form does the electron have in there?", one really did not know the answer. One was starting in the dark.

Well, I think here is one point of investigation. Is it a vast over-simplification to say that nuclear physics up to 1932 is essentially an experimental science? Certain concepts were introduced by great theorists like Pauli, but they were essentially phenomenological concepts, like the magnetic moment and spin, and then all that one did was to measure things. Can one say that Rutherford measured the gamma-ray spectrum, but there was not the slightest theoretical idea? I doubt it, but this is the point. This is an oversimplification. I would like to put it up for discussion and study.

Wheeler: You would include alpha decay as a very important theoretical development.

Weisskopf: Yes. When was Gamow's paper? [2]

Wheeler: '28, '29.

Weisskopf: Now, this is clearly a theoretical idea. I think that is very important.

2. G. Gamow, "Zur Quantentheorie des Atomkernes," *Zeitschrift für Physik, 51,* 204 (August 1928).

Wheeler: It is one of the most important theoretical approaches one had until '32 really.

Weisskopf: Yes, and here already was the idea of some kind of potential box with something in it, namely, alpha particles.

Morrison: The Gamow paper was proof of quantum mechanics, but the potential they already knew in '23 and '24. They had the size of the nucleus and the strong forces.

Weisskopf: So one already had this idea that some part of quantum mechanics is valid in there.

Wheeler: And then one had the Cockcroft-Walton work, which was really led to by this penetration idea. It came from the alpha decay; the alpha penetration led to the proton penetration which led to the Cockcroft-Walton work.

Weisskopf: The question is, for example, whether the preaccelerator nuclear experiments, like the first one of Rutherford's, had already stimulated discussions of the penetration of barriers.

Wilson: But that's a fact. Somebody suggested that protons would penetrate the barrier. Who was that person?

Wheeler: Well, Breit made an important contribution here. I do not know whether he was first, but he certainly was a leader.

Wilson: It was that suggestion that protons would be—what?—four times better than alphas just in penetration, which started everybody making proton machines.

Weisskopf: Here I would also like to ask my colleagues . . . This remark which is quoted [study paper, p. 252] about Bohr's having said: "Just as the account of those aspects of atomic constitution essential for the explanation of the ordinary physical and chemical properties of matter implies a renunciation of the classical ideal of causality, the features of atomic stability, still deeper-lying, responsible for the existence and the properties of atomic nuclei, may force us to renounce the very idea of energy balance."[3] My recollection is—and I might be completely wrong—that this was said because of the beta decay and not nuclear structure.

Segrè: Beta decay.

Weisskopf: Now, if we are right, Emilio, then I think that must be important in the interests of Niels Bohr's reputation. The point is this: It was obviously very exciting to everybody that beta decay seemingly does not conserve energy, although Pauli had mentioned the neutrino at an extremely early date.[4] But it was diffused very slowly and was not much believed. And Bohr was about to give up the energy law for the statistical validity of energy conservation. I think that these remarks here refer to that problem and not to the problem of nuclear structure.

Segrè: In 1929, there was already, though, a very big crisis on the even-odd number of particles in the nucleus from the Raman effect of nitrogen done by Rasetti.

Weisskopf: The Statistical Problem.

Wheeler: That was Ehrenfest and Oppenheimer.[5] Ehrenfest was visiting in Caltech.

Weisskopf: That was, of course, one of the things which was discussed at that time in Leipzig.

Segrè: Of course that was really something that was related to the neutron.

3. Niels Bohr, "Chemistry and the Quantum Theory of Atomic Constitution," The Faraday Lecture of 1930, *J. Chem. Soc.*, 379-383 (1932, Part 1).
4. C.S. Wu, "The Neutrino" in *Theoretical Physics in the Twentieth Century – A Memorial Volume to Wolfgang Pauli,* Fierz, M; Weisskopf, V.F., eds. (New York: Interscience, 1960).
5. P. Ehrenfest, R. Oppenheimer, "Note on the Statistics of Nuclei," *Phys. Rev., 37,* 333-338 (December 1930).

Weisskopf: Yes, that brings up a point. I wonder how much the idea of the neutron was already preformed. Had anybody predicted the neutron?

Segrè: Rutherford.[6]

Weisskopf: In what year?

Telegdi: 1920 was Rutherford's address.

Weisskopf: What did he say there?

Morrison: There should be a condensed hydrogen.

Weisskopf: But on what grounds did he say that? It was not on experimental grounds.

Morrison: It was on special grounds, analogical grounds.

Weisskopf: It was just intuition.

Segrè: At Cornell, Chadwick presented a paper[7] in which he told some of the experiments that he and Rutherford and others did to discover the neutron from 1920 [on].

Weisskopf: They really looked for something then?

Segrè: Well, they made a discharge in hydrogen to fire the electron into the proton, thinking it would make a neutron. They tried to find the neutron, but they didn't find it then [1920-1923].

Weisskopf: Well, all the more, I think, one has to be cautious about saying nuclear physics started in 1932.

6. Rutherford in his Bakerian Lecture of 1920, *Proc. Roy. Soc., A, 97,* 374 (1920).

7. J. Chadwick, "Some Personal Notes on the Search for the Neutron," *10th International Congress of the History of Science,* Ithaca, 1962, 159-162 (Paris: Hermann, 1964); J. Chadwick, "The Existence of the Neutron," *Proc. Roy. Soc., A, 137,* 229 (1932).

Holton: Do you mean to imply here that in further research, and also perhaps now, during the day, we should move this arbitrary date far back? Is there something that you are trying to say in terms of actual action . . .?

Weisskopf: I do not believe that anybody actually intended to study nuclear physics only from 1932 on. However, I get the impression from what is said—not only in this document [study paper], but from what many people say—that nuclear physics essentially started in 1932. And I just wanted to be sure that it is understood that this is not so.

Merton: Mr. Weisskopf, may I ask a question of intent? Are you also raising a question of whether a question of that sort—to say that nuclear physics was born in any given year—has any meaning?

Segrè: It was born in 1911[8] as America was discovered in 1492.

Weisskopf: Look, I could even defend, if I had to, that nuclear physics was born, but I don't know dates, as I said. When was the first Rutherford experiment with nitrogen?

Telegdi: 1919.[9]

Weisskopf: Well, I think that's really what we understand by nuclear physics. If we say that nuclear physics is the science of the structure, of the composition, of the nucleus—then that is the date.

Wilson: No, [it should date from the actual discovery of] the nucleus. It could have existed or not existed.

Weisskopf: Well, now, I would like to go to slightly different things. Who were the main forces, the main producers, of ideas at that time, 1932 plus or minus?

8. E. Rutherford, "The Scattering of Alpha and Beta Particles by Matter and the Structure of the Atom," *Phil. Mag.* 6th series, *21,* 669 (April 1911).
9. E. Rutherford, "Collision of A Particles with Light Atoms, Part IV. An Anomalous Effect in Nitrogen," *Phil. Mag.* 6th series, *37*, 581 (April 1919).

Now, here I would ask the following questions: To my mind the role of Heisenberg is not quite clear. I believe that Heisenberg must have had a tremendous role in all this. A specific question is this: The moment the neutron was discovered, and maybe even before, the moment the neutron was discussed, I would say, the problem of nuclear force must have come up. It must have come up with Joliot clearly, because, you see, something must keep those things together. But the problem of nuclear force in a more realistic analysis in the present sense, must have started when the neutron had become more and more of a reality. Now, who was the first to speak of the force and to worry about the nuclear force? I believe it was Heisenberg, but I am not sure.

Wheeler: In his paper [10] he spoke about neutrinos being exchanged. I am so interested in your having talked about the swimming pool incident, because I never had understood why the neutrino would be dragged in when it is so weak. But this gives good background for it, because he did talk of neutrinos being exchanged just in the way that one talks of electrons being exchanged in molecules in that great paper of his in 1932. But I really cannot understand why people today in books on nuclear physics do not refer to that paper of his more impressively. I suspect it is because we all now take so for granted the idea that neutrons and protons are in nuclei, and we take so for granted the idea of exchange forces, that we do not realize somebody had actually to write those two ideas down on paper.

Weisskopf: Of course Heisenberg introduced the exchange force idea. And now my question would be: What was there before? Were there other speculations about nuclear forces, or were these the first?

Telegdi: Besides Heisenberg there also were others independently among the Russians, like Ivanenko.

Segrè: I think that it was very much in the air, as soon as one had neutrons, the forces and so on. First of all, Ivanenko, and who was the other?

10. Heisenberg, "Structure of Atomic Nuclei," *Zeits.f.Phys., 77*, 1-11 (1932).

Telegdi: Tamm.

Segrè: Tamm and Ivanenko, Majorana, Heisenberg: they all did it independently at the same time, and I would not be surprised if you found another half dozen. And since helium had been treated—you see, with the exchange forces of the electrons which explains the singlet-triplet separation—everybody had that model.

Wheeler: But one had molecular physics too. At this time there was the work of Heitler and London.

Morrison: In Gamow's paper [11] it's already clear that there are non-Coulombic attractive forces that are very important in the universe. There is a footnote to Frenkel, Rutherford and Enskog,[12] all in '26 and 27.

Weisskopf: It is interesting to show that nuclear forces really go back much further. The deuteron, of course, should be emphasized as a very important point. The deuteron, the hydrogen atom of nuclear physics, after all, was discovered also in 1932. Here is a very important point for the development of the field: the paper[13] by Wigner about the infinitely short-range nuclear force, the theory of the deuteron. I might overestimate this paper because of my own personal experience, as for me this was the great thing. I am a Wigner student, of course, and I was very much impressed by Wigner. But as I remember my dates at the time, this paper by Wigner, on calculating the deuteron, was for me at least the beginning of nuclear physics in the calculating sense. From then on, for a theorist who wanted to calculate something, nuclear physics was the field to get into. As I said, I am not sure whether this is subjective or not.

Now, the discussion paper, I think, overdoes the point a little, that from 1932 on it was clear that quantum mechanics is valid in the nucleus. I feel that at

11. G. Gamow, *op. cit.*
12. J. Frenkel, "Die Electrodynamik des rotierenden Elektrons," *Zeits.f.Phys., 37,* 243 (1926); D. Enskog, "Das Bohrsche Magneton und die Radioaktivitaet," *Zeits.f.Phys., 45,* 852 (1927).
13. E. Wigner, "On the Mass Defect of Helium," *Phys. Rev., 43,* 252 (February 1933).

that time we did not doubt very much that quantum mechanics was valid in the nucleus. This is why I was also relating to Bohr here. I do not believe anybody thought at that time that quantum mechanics had to be revised. For such things as beta decay, yes, but not for the structure of the nucleus. This is my feeling.

Wilson: Still you would not have been surprised if quantum mechanics had not been valid in that region?

Weisskopf: You know, that was in the heyday of quantum mechanics. When you just win a battle, you believe that your weapons are very good.

Wilson: Maybe so, but no experiment would have surprised you. On the contrary, you would have immediately conformed to it.

Weisskopf: I am afraid you think that we were better than we were. That happened in 1931, and I was still at Göttingen. And then came Otto Stern from Hamburg to give a seminar. I think he already knew the magnetic moment of the proton but he did not want to tell it yet. He went around and took bets. He went to one theorist after another and said, "What do you think is the magnetic moment of the proton?" And everybody—at least the people I heard there, the big shots like Born and so on—said, "But obviously it's 1 magneton." Obviously, because it was just the heyday of the Dirac equation. "What else should it be?" And he grinned, of course, and took down the names, you see. Then he came back grinning in a few weeks announcing it's 2.7.

Segrè: Well, you see, Pauli, who was in Hamburg at that time, told him, "Why do you waste your time? Why should you do these experiments?"

Weisskopf: This incident might just partially answer your question about our confidence in those brand-new formulae.

Now, to proceed to beta decay. I think beta decay is also worth a close study by historians, because it developed rather in a different way just because it was always full of riddles—true and false—I mean experimental riddles. Some of them turned out to be experimental errors. First of all, that the energy law was not conserved led Bohr to strange ideas about statistical conservation.

Second, Fermi wrote a fantastic paper,[14] which I think stands out as a monument to Fermi's intuition. He made the simplest possible *Ansatz* for the beta decay. Then came years and years where the experimenters did not find the right results for the spectrum. Uhlenbeck and Konopinski developed a theory[15] with a derivative in it in order to fit the wrong experiment.

This is an example of how original intuition turned out to be right. But beta decay, with Fermi's idea, stands apart from the rest of nuclear physics because it is the creation of particles. It is the first example of modern field theory. We were completely unaccustomed at that time to think in terms of creation, though perhaps to say beginning of field theory would be an exaggeration because there were other papers, Heisenberg's and others. But this idea of creating particles comes from there, and beta decay was always shrouded in mystery. This is, of course, the fundamental side of beta decay.

The other side of it, the use of beta decay for studying matrix elements and transition probability, is another matter. That is ordinary quantum mechanics.

Well, I am not doing this systematically, but I would now like to say a few words about the attraction of the field of nuclear physics, again, referring to this paper [study paper]. I don't quite understand why it should be questioned because it is so obvious: that the subject of the nuclear structure and of nuclear particles was attractive, since they were obviously in the forefront. I mean, physics went that way because here was something that we did not understand and we had to understand it. The fact that the neutron was an exciting discovery, that the anomalous magnetic moment of the proton was an exciting discovery, as nuclear forces—this I don't believe in the tradition of physics needs any explanation. Obviously, this was, at that time, *the* thing for

14. E. Fermi, "Versuch einer Theorie der B-Strahlen," *Zeits.f.Phys., 88,* 161 (1934).
15. G.E. Uhlenbeck and E. J. Konopinski, "Higher Order Derivatives in the Interaction Ansatz of the Fermi Theory," *Phys. Rev., 48,* 107 (July 1935).

any physicist in the forefront of the development. I do not believe this is a problem.

Smith: You leave out one thing here. There always are lots of fundamental exciting, important things, But at this time it was possible to do something about this particular area.

Weisskopf: I think you have a point, of course. Maybe that is the thing that should be investigated. It is true that at that time when the neutron was discovered you could do something about it. Nevertheless, I believe the same important people who did something about it at the beginning had worried about it before. In other words, they did not change to the field because something could be done there. However, this statement might not be quite true. This is more of a question than a statement. Let me tell you again from a little personal experience. Shortly after '32, maybe rather in '34, quantum electrodynamics got stuck. Why do I say this? Well, there was a rather lively development in electrodynamics after the Dirac equation, the discovery of the positron, the pair theory—pair creation, the vacuum polarization. All this had a tremendous development, and many of the leading people, at the same time, who were in nuclear physics, were worrying about that. Then about 1934 electrodynamics got stuck. The infinities were there, and one could not do anything.

Now, here to my mind is a very interesting problem: Why did it get stuck? There is no reason in the world that the discovery of the renormalization of Schwinger, Feynman, and so on, couldn't have been done at that time. In fact, everything was prepared. I know it very well myself, because I was just too stupid. If I had just followed a few of my own papers, I could have gotten it at that time. Intellectually everything was prepared at that time.

Telegdi: It wasn't experimentally.

Weisskopf: Well, I think there are two reasons in my mind. One is the pressure of the Retherford-Lamb experiment which came later. But I must say, Val, I am not so sure about it, because the problem was there before.

Morrison: E.J. Williams had almost the same answer. Much earlier.

Telegdi: That was '38. I think there was not the pressure to really decide it experimentally.

Morrison: But the other thing comes out in Heitler's book very well. It was very strong in Berkeley, I know. Through Robert Oppenheimer it was very strong. Namely, the cascade process was not understood. Heitler's first edition tells all about it. There's a breakdown in electrodynamics of 137 mc^2 very probably because you see these multiple particles coming out.

Weisskopf: When was the paper by Heitler and Bethe about the cascade?

Morrison: No, no. The cascades are Carlson and Oppenheimer[16] in '37. And then Landau and Rumer '38.[17]

Weisskopf: Fine. This is certainly an explanation. But I think there is also another explanation: that there was another field of activity.

Telegdi: I think that what Phil [Morrison] is saying is very true, because we did things and found great surprises: electrodynamics works above 100 MeV.

Wheeler: I think, though, that these fields of physics were not as separate as they appear on the surface. I would like to say just a few words about the development of quantum electrodynamics, because there was the work at that time to study the scatterings of high-energy gamma rays by the nucleus. Of course, Delbruck had been working with Heisenberg trying to calculate from the pair theory the likelihood that a gamma ray would be scattered by the nucleus, but poor Delbruck was driven nearly crazy by this being the most complicated calculation anybody had ever tried till that time. He nearly did

16. Carlson, Oppenheimer, "On Multiplicative Showers," *Phys. Rev., 51,* 220 (February 1937).

17. W. Heitler, H. J. Bhabha, "Theory of Cosmic Ray Showers," *Proc. Roy. Soc., A, 159,* 432 (April 1937). L. Landau, G. Rumer, "Cascade Theory of Electronic Showers," *Proc. Roy. Soc., A, 166,* 213 (May 1938).

come through with the answer. But the experimentalists were working, and there were Gray and Tarrant and there was Jacobsen and there was Meitner: four different people working on this scattering of 2.6 MeV gamma rays by lead. And the amazing thing is that all of them got a scattering far larger than anybody could explain in any reasonable way. So it looked as if there was some new and unexplained interaction of gamma rays with the nucleus, and this really attracted a lot of attention. This was 1934, 1935, in that period.

Then Williams got quite interested in this, and he asked very important questions about the possibility that this was really due to some kind of multiple process. That is to say, pairs were produced in the lead; the electrons were scattered, and then they reradiated, and that this was really—through this complicated chain of events—a miniature cascade. You were getting the scattering. This, in fact, worked out and people lost their interest in this scattering by the nucleus as giving anything truly fundamental.

This was the time, of course of something else that was very relevant—namely, the work of Rossi and Gentner and others on the penetration of cosmic rays through lead. How in the world could a negatively-charged particle get through lead? A positively-charged particle—that one could understand. It could be a proton. But a negatively-charged particle: how could it get through? Because pair theory would be violated. Of course, at this time one had the 1934 Bethe-Heitler calculations[18] which said that you could not get through. But those were complicated calculations, and who could believe that one really knew the facts? And then along came the arguments of Williams and Weizsäcker, in which Bohr took a very active part, saying that really all that it depended upon was the Compton effect. And from that you could predict by relativistic transformation that a high-energy electron just had to radiate like everything, and a high-energy electron just could not get through the lead. This really prepared the way for Carl Anderson's discovery of the meson. Because if you look at Carl Anderson's paper[19] on the meson, you discover that it is the penetrating power that he has measured, the energy

18. H. Bethe, W. Heitler, "On the Stopping of Fast Particles and On the Creation of Positive Electrons," *Proc. Roy. Soc., A, 146,* 856 (1934).
19. C.D. Anderson, S. Neddermeyer, "Cosmic Ray Particles of Intermediate Mass," *Phys. Rev., 54,* 88 (July 1938).

loss, and without this background of the study of electrodynamics, one would never have really had the tools to discover the meson. So it is very interesting how this electrodynamics interacted with the nuclear physics.

Weisskopf: Oh, yes. I would say so, if one makes this artificial distinction not so much with nuclear physics, but with elementary particle physics. I would like to say a few words about this later.

Wilson: There was another important part about the relationship between the theory and the experiments, and that was Williams' explanation in very simple terms of what was happening; so that an experimentalist could understand it—and, in addition, would believe it. Because at that time, at least to me, it was something that I did not believe particularly.

Weisskopf: There was another thing when I look a little further at the development of nuclear physics in the '30's, which I was always struck with, and that is a certain change in attitude towards the field. What I really mean is that the theorists turned to phenomenology. The point is that more and more facts, came up not only due to the development of accelerators, but due simply to more nuclear experiments with natural sources. More and more facts became known, and then theorists were just forced by these facts to explain them and to try not to go after the fundamental question—what is a nuclear force?—but to say, "Now, imagine there is some kind of nuclear force. How can we understand that kind of thing which we see?" Well, I would call it a turn to phenomenological physics. And I think Bohr's compound nucleus is a very typical example of this. Here one tried to understand a phenomenon not by understanding the details. It is like understanding the solid state by assuming some experimental constants like lattice systems and elasticity and the rest, and it is a kind of phenomenology. This is a change in the work of some people, of most of these nuclear physics people, at that time when Bohr introduced the compound nucleus, which was '36, I believe.

Then with Bohr's compound nucleus and its influence of course, there came the evaporation theory and all this kind of thing; the Breit-Wigner formula . . .

Segrè: That was before. Breit-Wigner was before the compound nucleus.

Weisskopf: Yes, but it was after Fermi. When was Fermi's experiment?

Segrè: In '35. He showed the resonances.

Weisskopf: Then came Breit-Wigner and then the compound nucleus. So I think the Breit-Wigner formula is perhaps the first example of what I want to say: where phenomenologically, with some fundamental physics concepts, one tried to describe the phenomenon. And the compound nucleus afterwards.

Now, this I think has a psychological effect on many people. To put it again, this is a question; not a statement. I believe that some people were attracted by this, and some people were, in fact, repelled by it.

Wheeler: Right. So many people said, "Well, give me the Hamiltonian, then I'll compute. If you haven't any Hamiltonian, I'm not interested." And there was no Hamiltonian.

Weisskopf: That's right. And the fundamentalists were sort of, at that time, pushed away. I believe that this is one of the reasons for the split between nuclear physics and elementary particle physics. I would say one perhaps could put this time as the beginning of that split.

Holton: Would you elaborate on the word, fundamentalist?

Weisskopf: Oh, I would have a hard time. I cannot really define it; I can only give examples. Heisenberg was a fundamentalist. There are people who do not like that kind of thing—[do not] like, for example, the compound nucleus idea of evaporation models. It wasn't that they disliked it, but they didn't work in that field. To them what is really interesting is: What is the nuclear field? The Yukawa theory. That is real physics and that is the kind of thing we have to go for, for the basic electrodynamics. This is what I call a fundamentalist. Sometimes I call it intensivist, but that is also wrong. It is for you to invent the names to describe things; it is for the sociologists of science.

Again, I am very unsystematic. Let me now go over to the latter part of the original division [Study Paper] —the sociological aspects. Here questions were raised in this paper. I have already expressed my views on one question—namely, why were people interested in nuclear physics? I think this is, to my mind, an obvious question. But there is, of course, the question: Why did nuclear physics develop the way it did? For example, why at that time did the center of nuclear research slowly go to the United States? But let us state, if you talk about the 1930's, that it was not completely the United States; by no means, in fact. The Fermi experiments were made in Rome, and a lot of the cosmic ray experiments were also made in Europe; and much of the theory we are talking about was made in England and Europe. Of course, Germany at that time was a black spot on the map—or a brown spot.

There is now the very interesting problem of how far the so-called technologically advanced state of the United States at that time influenced the development of nuclear physics? There is no question that the accelerator played an enormous part. The year 1932 is not only the year of the discovery of the neutron; it was the year of the first accelerator. And here are a number of questions which I am extremely interested in, and I am sure that Stan Livingston will be a very good man to answer these. It is tempting to say from our present point of view that the development of the accelerator, which was certainly done in America, is due to the fact that America had more money and more technological support, but I wonder whether that is right or whether it is the speed in decision-making which is faster here and [the fact that] the universities are better supported. This is all true now. But was it true at that time?

One can ask the question: Why is it that this accelerator development, which was started in England with Cockcroft and Walton, was then essentially taken over by the United States? Now, if one discards the rough trivial explanation which I mentioned before, the other thing, of course, is the personality question. The personality question comes up all the time, and there was only one E.O. Lawrence, and that might have something to do with it. But all this is in the form of a question to Stan Livingston.

Livingston: Well, do you want me to give a little story about this? I actually took the time to write up a study paper of one page that tried to express my views on how the accelerator entered into the field. I would make first the usual disclaimer: I am a specialist and therefore an enthusiast and therefore biased.

I start with a point: The energy of particle accelerators increased by enormously large steps, that is, the ratios from one to the next were great. That I think is *the* major point which made for the rapid progress of nuclear physics. Now, that may seem exaggerated, but it is the theme I am going to try to prove. I even have these ratios illustrated, and I went back and searched the literature to find the ratio of voltages, the jump that the cyclotron brought over the Cockcroft-Walton. If you look at it in 1934, that jump was a factor of 10 in energy.

Cockcroft-Walton never got above 700 KeV in those early days, and the cyclotron was about ten times higher by 1934. But by 1940 the factor was up to 20. In other words, it had jumped by a factor of 20. Now, let me just quickly run through a few of these ratios. Betatron to the electrostatic generator, it was roughly 5 in '42, roughly 10 in '45. A factor of 10 above. By '45 the betatron had reached an energy 10 times larger than the electrostatic generator at that date. So similarly, the synchrocyclotron over the cyclotron: a factor of 10 in '46, a factor of 20 in '48. The electron synchrotron over the betatron—this was a smaller jump, a factor of 4 by 1946. The proton synchrotron over the synchrocyclotron, a factor of 7 by '52; that reached a factor of 12 by '54. That was the cosmotron first, then the bevatron, and then eventually the big one such as the ZGS at Argonne.

Now, the AG idea came in later and didn't have any significance until the early '60s. But there, the AG proton synchrotoron raised it over the proton synchrotron by a factor of 5 by 1961, but because the ZGS at Argonne came along, it dropped to a factor of 3 by 1967—by now. The AG electron synchrotron was boosted over the pre-AG type by a factor of 5 by our machine here at Cambridge. That was by '62. By now it is about a factor of 6. The latest one, which is a little different, is the electron linac at Stanford, which is going up over our electron synchrotron by a factor of 3, which is happening this year. It [began] last year.

Now, I am going to use these large ratios to prove something; namely that each step was based on a new principle of acceleration or a new technical feature and originated by one—or occasionally by a few—accelerator scientists, who were physicists. They saw the need for high energies, but did not have any grand visions of a new field of nuclear physics developing. But they were specialists, and they developed a class of specialists who were accelerator designers. These were the people that conceived and developed these ideas, and they had their own, almost separate, motivations for developing more and more energy. These concepts led to upward steps in energies, and they originated with these accelerator designers, not with the research scientists and the nuclear physicists of the day. However, these higher energies were of course appreciated and used immediately by the research scientists. But the point I make then, is that for a reason which is almost external to the general development of nuclear physics, the team of particle accelerator builders developed their own little special field by great big steps. And those were the stepping stones by which the experimentalist then picked up and made the fast progress.

This is basically my thinking: that progress came because these large steps opened up new phenomena. Almost every one of these major upward steps in energies led to opening up a new area of scientific interest, a new set of phenomena, which at the start any competent scientist with the opportunity to use could exploit to find new things. I am sorry to say a few Nobel Prizes came that way, as by-products of somebody else's work. Those, too, were immediate things. But still, later on, after these exploratory surveys were largely completed, the research scientist came into his own and really began to exploit and develop the field with more sophisticated and significant experiments. But in general the early exploratory experiments were easy.

Smith: Do you think it is right to say that this development was external to nuclear physics and that part of the motive of building the things was enjoyment?

Livingston: The enjoyment was in working with the research scientist who obviously wanted this and appreciated it and wanted to do it. But, to me, it was sort of a different thing. I do not think Ernest Lawrence was particularly impressed with the growth of an enormous area of nuclear physics, but he thoroughly enjoyed the technical success.

Weisskopf: I do not quite follow this. Is this really so? I mean you and Lawrence certainly had most of the fun and most of your intellectual activity was actually in the accelerators. But I cannot imagine that you would have done it if it hadn't opened up a whole new world for science.

Segrè: No, I think Livingston was very right. I remember very clearly the atmosphere at Berkeley. When I arrived there it was as late as '38. Remember also that the neutron, artificial radioactivity, fission—all these things—none of them came out of the machines. Nothing came out of the machines.

Morrison: The meson?

Segrè: Not even the meson.

Telegdi: But they could have made mesons.

Segrè: But in particle physics it is a different story. Let us stick to nuclei, because the role of the accelerator in particle physics is a very different story.

Weisskopf: The question really is: Is this a psychological question or not? Was this background noise, to say the least, important for you or not?

Tuve: Well, I'd belong to the Italian Primitives rather than the Impressionists. I would say that the motivation was entirely a breakdown of Maxwell's equations in the structure of the nucleus. As far as our Carnegie Institution group was concerned, this was the motivation from the beginning of 1926. The work started in 1927. I can't assess the importance of jumps to factors of 10 in voltage after, say, 1933 or '4, because most of the basic discoveries were made by then. Most of the basic clues to this area of physics had been established. This is why in this write-up I was a little bit bothered by the beginning of nuclear physics in 1932. Actually, I remember in my student days in 1920 and '21 what the basic excitement was in physics. You must remember that there were very few research physicists at that time. So the fact that there were not very many nuclear physicists was simply a natural consequence of the fact that there were not many research physicists. But Tate and Swann, [A.] Zeleny and [H. A.] Erikson at Minnesota in the 1920s—their real interest was in radioactivity and ionizing radiations. This was called atomic physics at that time. And it became quite clear following up Rutherford's artificial disintegration, as it was called in the middle and late '20s, that the exciting thing was that you could really hope to explore the breakdown of the ordinary Maxwell's equations by the time you got into the nucleus. At that time we were not talking neutrons. This was not a forgotten or an ignored idea, but it just did not seem to work. It seemed to be protons and electrons as far as you could figure out. It was quite clear that you could not build a nucleus, a positive nucleus, that would stick together unless there was a breakdown in Maxwell's equations. And the way to get at it was clearly by essentially scattering experiments and perhaps disruptive experiments ultimately. So the first key I believe was Condon-Gurney-Gamow. It grew out of atomic physics with the idea of potential wells and penetrations, following Schrödinger and Heisenberg and Dirac in the middle and late '20s. It was quite clear that this was the new chance in physics.

Of course, all of us youngsters were, I believe, extremists. We always wanted to go to extreme of temperature, extreme of pressure, extreme of voltage, extreme of vacuum, extreme of something or other. So actually I was going over to Rutherford's lab in late 1926. But Fleming at the Carnegie where I had been with Breit said, "Can you finish this in one year?" I said, "I want to go and put a million volts on a vacuum tube." He said, "Can you do that in one year?" I said, "Lord, no." He said, "Well, why do you go to Cambridge? You've got to quit at the end of the year on the fellowship. Why don't you stay here and do it?"

So it was as specific as that back in the middle '20s. And I think this follows on through. The guiding idea was: really how do you get at such a fundamental breakdown as the repulsion of positive charges? Most of you know the kind of disappointing stresses that we went through trying to get voltage sources. Tesla coils were doing 6 [kilovolts], but we did what we could with a few dollars, and then succeeded, if you'll remember, in getting penetrating gamma rays and measuring their penetration coefficients and beta rays.

Wheeler: You started to build this before the Cockcroft-Walton?

Tuve: Oh, yes, sure. In fact, it was December 1930, January 1931, that we presented, at the AAAS in Cleveland, pictures of protons with a range of a centimeter and a half in a cloud chamber and measurements and pictures of beta rays bent in the magnetic field showing their velocities were up over a million volts. Measurements of the absorption coefficient of gamma rays had revealed that they were also in this range. For the fun of it, I reviewed our old annual reports at the Carnegie from 1925 till 1930, and I find that even there expression was given to the fact that we wanted to have high voltage tubes in order to probe the nucleus. Actually, the words were not used quite this way: in order to study the breakdown of the laws of electricity and magnetism in such a small structure as the nucleus. But it was clear that a million volts on a proton was the order of magnitude required.

Wheeler: Where did your calculations come from? Was Breit active in discussions of this?

Tuve: Yes. Breit and I were together there. You see, he was at the Carnegie first. Well, you've got to remember there was a Depression; there were a whole lot of auxiliary, let's say, social, factors. But the fact is that being so irked by the short on-time of the high-voltage impulse from a Tesla coil, we found we actually stopped trying to use it in 1930 until we could get money for a steady high-voltage source.

Now, the calculations had been made by Condon-Gurney[20] and by Gamow.[21] We were familiar with them. In fact, we brought Gamow to Washington in, I suppose, 1934. I have not really called to mind just which year that was.

20. Gurney and Condon, *Phys. Rev. 33*, 127 (February 1929).
21. Gamow, *Zeits.f.Phys., 51*, 204 (1928).
    Gamow, *Zeits.f.Phys., 52*, 510 (1929).

But it was on account of his calculations of potential wells that we brought him over to join in our journal club. As I say, this goes back to the days of the Primitives. Maybe I am thinking of an art museum rather than the Fermi and Rome group. It was far from primitive there.

Morrison: Did you know about Cockcroft and Walton? Their first paper seems to have been in the '30s.[22] Their publication was about the same time.

Tuve: I suppose there was some vague knowledge. Sure, everybody recognized that we needed high voltages to get artificially accelerated particles in order to do this job. But, of course, I personally felt like a chump because I had known for a couple of years . . . It did not occur to us to try to run things. Remember, they ran up to only 2 and 100 kilovolts for their first disintegrations, and 3- and 400 was ample. Whereas in the beginning we kept driving for these very high voltages. A million volts was the goal.

Wilson: What date did you start doing experiments, though?

Tuve: I withdrew my application for a National Research Fellowship to go to England in the spring of 1926 in order to come to Carnegie for this express reason. You see, we had carried out the pulse radio experiments in the summer of '25 and shown that this was a way of studying the upper atmosphere. And Fleming wanted me to come back and join in that work, help it along, but I said no, I wanted to go to England to work with Rutherford. This is the reason I remember the dates.

Wilson: I meant that after you had built your tube, then when did you make your first experiment on the penetration of gamma rays?

Tuve: We gave the paper at the December 1930-January 1931 AAAS meeting. Actually they flattered us with the AAAS prize on January 3, 1931.

Weisskopf: Do I understand you to say that you have a different

22. Cockcroft and Walton, "Experiments with High Velocity Positive Ions," *Proc. Roy. Soc., 129,* 477 (1930).

interpretation of why you people had worked on an apparatus than what you heard from Stan?

Tuve: We were talking of different epochs. That is the reason I said I go back really to the prehistoric period when there was no history of nuclear physics except who was who. That is the major point I want to make: that nuclear physics was the main center of active interest from 1918 on.

Segrè: No, no. There was the development of quantum mechanics.

Tuve: Well, that happened to be something that fortunately worked out.

Wilson: That was the result of the Rutherford model of the atom which had a nucleus in the center of it.

Morrison: It was all called atomic physics.

Weisskopf: It seems to me here may be an interesting question of what really brought this wave of accelerators. I raised a question which I do not think you answered, namely: What was it that made you and Lawrence so successful? It was not the money support. I think you shook your head. Compared to Europe, you see.

Livingston: The state of the art was a big part. And it was the inventive ideas that Lawrence had that made it possible.

Weisskopf: I also asked the question: Why was it that the development of accelerators (I'm not speaking of after the war, but pre-war) was all in America with very few exceptions? Was it the personalities or was it because of much more money and industrial support?

Morrison: Also the Van de Graaff comes in here.

Weisskopf: Van de Graaff was also in America.

Livingston: I think that actually it was a sequence of separate individual cases of real capabilities and inventiveness: Lawrence, Van de Graaff; Tuve has been describing his role. This thing developed in several centers, and I think it was in a certain sense due to the ingenuity and technical skills that have been developed by the young scientists in this country, which may have been different—I am not sure whether they were different in type—from what was developing in Europe, but at least there was a great deal of technical skill. Most of us, for example, were radio hams and had taken apart Model T Fords, and we learned mechanics as youngsters. I think this had a great deal to do with being prepared to do the kind of technical thing that those of us who were specialists did.

Smith: Let us hear from Dr. Herb, who was mixed up with this.

Herb: I agree with Stan [Livingston] in regard to this. I think young boys growing up in this country tended to take to complicated apparatus. And I believe my interest in nuclear physics came after a talk by R. H. Fowler in 1928. It was a most inspiring talk I happened to hear and I got interested. He was from the Cavendish. He made nuclear physics sound very inspiring. I was most impressed. This was at Wisconsin. It was quite evident in his talking about the alpha spectra, what had been done at the Cavendish. A lot was known, and it was quite intriguing—a lot was unknown, of course. But this was my motivation. And then with Van de Graaff's development, first I started with another fellow attempting to put this in a vacuum. This was a most disillusioning experience. And when he left, I got at the job of putting it into a pressure tank, and this worked. With this I started a program of nuclear physics. I had the advantage of having Gregory Breit there, and naturally he was most interested in seeing more experiments in nuclear physics.

But with the Cockcroft-Walton machine having been developed in Europe, it was quite significant that practically all these machines were built in this country. I do not think the cyclotron or the electrostatic machine necessarily should take the higher role in regard to productivity. The Cockcroft-Walton machines just multiplied in this country. There were tremendous numbers, and they did a lot of the work. And I think the fact that Americans took to machines played a fairly big role. It certainly wasn't money because we didn't have money. My first machine I think cost about $250. I took an old tank—all old equipment—and Don Kerst wound the transformers for the power supply for the dome. He was a good transformer winder. Everything was made by hand. I think the most expensive purchase was a little air compressor.

Merton: Did you have any great doubts that you could build the machines that you had in mind? Or was it just applying techniques that you had learned earlier and that you had just assumed would work?

Herb: No. I certainly was not applying techniques I had learned earlier because these things had not been found.

Tuve: The vacuum tube problem was really uncertain when all this started.

Herb: All of us started in complete ignorance, and we just went ahead and tried it. I think one of the advantages was that it was extremely cheap and we could afford to make a lot of mistakes. I think Kerst and Parkinson and I probably went through 25 vacuum tubes quickly and finally progressed to one that held 400 kilovolts with a little less than 2 feet. But this was not planned. We just went ahead and made mistakes and profited from mistakes. I think the same was true at the Carnegie Institution, wasn't it?

Tuve: When we started, of course, [W.D.] Coolidge had managed to put 220 kilovolts on an x-ray tube. This was about 1926, about the same time. He had made one cascade tube which illustrated the principle. We punctured some very expensive tubes that we got from G.E. These were gifts, fortunately. But there was a real hurdle. You could not put more than 250,000 volts on a vacuum tube without puncture. This remained true for a couple of years, so this was really an important thing in the '30s, and at this time we had shown that you could really have these voltages. Don't forget, it was a little difficult to measure the voltage on a Tesla coil. The only thing I trusted was the fact we could bend the electrons and we could measure the range of the protons, and that was the reason for these measurements of penetrating power. To prove that you really had gotten over this apparently insurmountable hurdle of putting really high voltages, many times 250 kilovolts, on a single vacuum tube so you could have a free path, was another question.

Let me make just one comment. I think that the basic thing was that there was bound to be an unfolding of an enormous amount of new investigation and study once the basic things had been discovered. Not only, of course, quantum theory, which set the stage and gave us tools to work with, but the discovery of the neutron, artificial radioactivity, [and] artificial disintegration. Then it just simply insured that wherever there were people, who were qualified to work with their hands and put things together, there was bound to be activity. And I think there was considerably greater freedom to attack things in the United States, [due to] the temper of the boys who were used to working with gadgets, than in most other countries.

Herb: I think there is also the fact that in this country we probably were freer to do things. When I changed the tank over to a pressure tank, my major professor, who was a very fine fellow, did not want me to do this. I was just a second-year graduate student, I guess. I went ahead anyway and did it, and he did not mind. He had expressed his objection—that was enough. He went along with it fine. I think this probably would have been difficult in Europe.

Wilson: I think what is being said is that the development of these high energy accelerators was something that was typically American. It came because Americans who were not well-to-do had reached a level of education. Those who became professors of physics were sophisticated enough to appreciate that physics was an exciting field, but they had come from a severe background in which they committed themselves completely to doing whatever was necessary to get on with the work. Generally speaking, because of the poverty of the times, this required that they had to do all the work themselves. And so they did it.

Tuve: The amateur status was very important. They were accustomed to it from ham radios and from automobiles.

Wilson: Because we are a nation of tinkerers this was very much in our tradition. Yet these were sophisticated people. They could appreciate that physics was a field of great interest, of great opportunity.

Holton: But from our present point of view, I think there are several things which are going to have to be studied. One of them is whether there is a great difference here between the levels of poverty at which these experiments were being done. That is to say, we are all agreed that relatively little money went in, but I remember that in Chadwick's paper for the Cornell conference[23] he said he had worked for two years to find the neutron and all of these were failures. "I had wasted two years but I had wasted none of the laboratory's money," which I think at that point was a yearly budget of £2,000, plus or minus, for the experiments. What we are hearing here is that people in this country were willing to "waste" at least a little money; and that can be quite a crucial difference in the scale of the enterprise. A few hundred dollars for a Lawrence as against the pride of wasting none of the money can make a great difference; the willingness to go ahead and break a few expensive tubes even if they are gifts, but that may not be replaceable as gifts; the willingness to do some experiments with equipment which was not cleared for it and which may in fact not work afterwards for the initially intended purpose. These are acts of a kind of courage which I think makes the scale of operation qualitatively different.

And there is the point that we don't, after all, find this only in nuclear physics. In America there is a tradition of the Michelson and the Millikan kind of an approach, the joy of working with certain equipment on problems that may or may not be in phase with the state of the theory. And I remember most of all Mr. Bridgman, who in his laboratory, when I was a student there, was really driving with a single-minded passion to up this factor which Livingston speaks of by a factor of 2 to 5 to 10. That was his joy. And every year he gave a review paper in our colloquium series in which there would typically be 55 new measurements which theorists did not understand. It was really the joy of doing a field, sweeping through all the factors of 10 that one could do with one's equipment.

Smith: And the desire just to explore and find what was there. It was not only a question of enough desire but it was the thing to do.

23. *Proc. 10th International Congress of the History of Science,* Ithaca, 1962, 159-162 (Paris: Hermann, 1964).

Morrison: It was very much the same as Kapitza.

Livingston: There is one more aspect about the financial side of this, about accelerators, that I would like to point out, which is that almost each one of these new ideas, new conceptions of accelerators, led to a reduction in the unit cost. Although energy went up by large amounts, the cost did not go up by nearly as large amounts. And it was this apparent saving and the apparent achievements of the new great accelerators which helped get them accepted so that they could be built. Money could be raised because they were more efficient in terms of money.

Morrison: There's hope for pennies still.

Cohen: May I ask one question which arises here which interests me? I am sure that the tinkering and ham radio were very important factors. It is a hobby of mine to look for the ham radio behind a lot of physics, and Rutherford, of course, himself began in that. But are we to understand from this that in Europe, Italy, Germany, Austria, boys did not tinker with ham radios? I find that rather astonishing. Surely there must have been quite a lot of that.

Segrè: Yes, but not at the same time. The radio came in 1917. So if you take a boy of 10, 15 years old, he could not because there was no radio. Say, for instance, Fermi was a great maker of electric motors. I remember I had a Ruhmkorff coil and I had my discharge tubes, and I amused myself with them. So did many other people of my age.

Telegdi: I want to add that I know that Szilard invented some of these accelerators—I don't know which ones, maybe even several. He had patents on them. He tried to go to people like German scientists and General Electric and convince these people it would be a good thing in 1928, or whenever it was, to build an accelerator for nuclear physics. They had no interest whatsoever, and it must also be said that he was not the type who would sit down and tinker and build a gadget; so I think the fact that there were people who had ideas and were tinkerers at the same time is a very important aspect. And I think there is among the people who must have gone into physics at that time a difference in background between Europe and here.

Morrison: Could I ask a question? [C.C.] Lauritsen came to the radiological side where there was really economic desire, so to speak, for high-energy fields. Does that play any role in the nuclear accelerator development? Certainly there was an interaction.

Herb: Certainly Lauritsen's work was very important in nuclear physics. But I suppose it did not play a very big role in other machine developments, although it probably helped some.

Morrison: I meant the other way around. Did you at all think that any of these proton beams would be useful for radiology?

Herb: I was thinking entirely of nuclear physics.

Tuve: I suppose I knew Charlie [Lauritsen] in 1930. Of course that was simply a stepping stone. The Kellogg lab was built because of the interest in penetrating X-rays. But he had fortunately cleared a space in their high voltage testing room, and over in a corner he had a tube for physics. He used this to pay the laboratory's salaries and expenses, but he had the other one to pay the laboratory's debts. Crane and Lauritsen were busy working on tubes and moving toward higher voltages. They were mostly accelerating electrons, and things like that, but the aim was positive ions.

Livingston: Quite a few of the big accelerators were supported by medical foundation money and in the early days for X-rays, as you say. Later on, E.O. Lawrence's 60-inch [cyclotron] was supported for the neutron's [application to medical research].

At MIT the cyclotron was set up with money from a medical foundation, and this is the truth as most of you have said, with the urge to get high-energy X-rays. I know in 1933, in Berkeley, Lawrence couldn't get me an instructorship for the second year to pay my salary, so he got me a part-time job building an x-ray tube of Sloan's design in the San Francisco hospital. It operated at one and a quarter million volts. It was Dave Sloan's design of a resonance transformer x-ray tube, and it was sort of done with the left hand while we were trying to keep the cyclotron development going with the other. The stimulation of medical people wanting x-rays was a big part of the original development.

Tuve: Speaking of Europe, let me remind Telegdi. Do you recall Brasch and Lange subsidized by some Swiss power concern? I think they were trying to string a wire between mountain peaks in order to do nuclear physics.

Telegdi: One guy got killed.

Tuve: That was 1928. So there wasn't a lack of commercial interest. That was spectacular.

Weisskopf: Returning to this comparison between Europe and America, let us not forget that the physics community is a highly non-linear system. And a very small difference makes a lot of effect. It is of course ridiculous to say that the European boys did not tinker. Of course they did. But a small difference has other logical effects, I think.

Herb: So in the laboratory, things were freer and easier here. The professor did not dictate in detail what people around did.

Holton: But look at Rutherford's handling of the Cavendish: nobody comes in at night.

Tuve: Yes, I am sure this was a major factor, in my opinion. Herb is right.

Herb: Certainly I could not have done it if I had been closely supervised.

Tuve: There was an invitation to contribute. That was essentially the attitude.

Kuhn: It seems to me necessary to underline what was really implicit in what Dr. Holton said before simply as another factor. Of course the American tradition had been largely an experimental tradition and an apparatus tradition to the extent that it had existed at all. However, what there was in the main

European centers and was not in America until the late '20s was a reasonable sort of experimental tradition. So that this was the first time, I would suppose, that the American tradition as an experimental tradition really could hitch on to a hot forefront field without the initial backing of the sort of theoretical work which was in Europe playing hand in hand with experimental work but still involved in cleaning up quantum mechanics. And I would suppose that this gave American experimental physics in the new field something of a lead, which it also had the potentiality to exploit.

Wilson: Can I just say one thing about the matter of the role of the individual. In this case it is true and you can point to the individual. This field would have developed completely differently without Ernest Lawrence. He was a man of overwhelming ambition who had a vision of how the field might develop. You can compare his laboratory to the great astronomical laboratories of the time. If he had not existed, we would have seen the developments that did occur in Wisconsin and Washington, but those would have occurred more in the European tradition. After all, some things were being done in Europe. The Americans would have gone along at about that level. I doubt that the Radiation Laboratory would have come up to the level of a great laboratory before the war without Lawrence. We would have seen that physics would have developed subsequently in just a completely different manner – an order of magnitude different manner.

Smith: I think I will ask Professor Weisskopf to bring up the one or two points that he still has hanging. Incidentally, I am assuming that nobody objects to this somewhat unstructured discussion that we are having. It seems to me that it is this kind of thing that will bring out the points that we have not already thought about. From the standpoint of the historian, I am sure that the accidental revealing statements have been much more valuable than the formal statements that have been made by the physicists.

Weisskopf: I now come to a point which is complementary to the one we discussed before. That is what one falsely and popularly may call the fusion of European theory with American experimental know-how. I am formulating it this way so that you will know what I am talking about. To begin with,

I object to it because there was a strong theoretical tradition in the United States--Oppenheimer, Breit, Rabi, Van Vleck.

Kuhn: Not so much the tradition as a critical mass at this time.

Weisskopf: I again come back to my statement that I believe in all these things that one must not forget that a scientific community is a highly non-linear system, and very small differences have very large effects. And the differences, of course, were not so small. First of all, it is always disagreeable for me to talk about national or even continental traditions in science. There is of course something in the idea that there is an American tradition of tinkering. But in another way, of course, the whole idea is somehow against the principle of the scientific world. Science, and particularly physics, is very strongly international and has always been influenced by both sides. So that, if there is any international community, it is the scientific community where those national differences are reduced to a minimum. But of course they are still there.

Somehow discussion of, is it European influence or is it American influence, is on the level of a question like: "Has the East Coast or the West Coast played a more important role in the development of nuclear theory?" This is also a silly question.

It seems to me, again to come back to the personality angle, that this is an important question for discussion. The influence of the refugee theorist in Europe and America is mainly, perhaps even exclusively, an effect of personality. Take Bethe, for example. I would like to take Bethe in many respects as an important example. Bethe and Wigner played an important role in the development of physics. Now, let us not forget that moving Bethe and Wigner from Europe to America is minus Bethe-Wigner in Europe and plus Bethe-Wigner in America. That means delta is 2n not n. If you talk only about those two, who might have come over because of completely different reasons, it would have probably made a decisive difference. It would have had a decisive effect on any physics which those two people joined. That is one thing.

The second thing, of course, is not only the minus of Wigner and Bethe; it is the minus of the whole scientific tradition in Germany. And Germany is not all of Europe—let us not forget this. England and France and Scandinavia are very important in nuclear physics. However, it is minus Germany and also minus, to a great extent, Italy in that certain period of the '30s that had this effect.

Anyhow, despite all this, it seems to me that the influence of Wigner and Bethe is certainly not negligible. In fact, I think it had a major effect, and I would like to say a few words about it. The "Bethe Bible," of course, is part of it. Rather, it is a symptom of this essential effect. One of the things which Bethe, perhaps more than Wigner, has brought over—and, again, this is not a European trait, it is just Bethe—is the theory-experiment relation. It is certainly wrong to say that a close collaboration between theorists and experimentalists is a European tradition. It is not. In fact, it is the opposite. However, again, there is no question that Bethe brought this into the United States. Perhaps it was already here. Breit had it. Rabi in a way also fostered this, although Rabi is not a theorist, but he fostered it. Oppenheimer probably had it to a lesser extent.

Wilson: But it was here and at an important level because it involved Ernest Lawrence.

Weisskopf: So we have developed the important point that this essential element in the development of nuclear physics—namely, the close collaboration of theory and experiment—certainly does not come from a transplantation of the European tradition. On the contrary, I would say that Bethe was very essential. I must not omit Fermi in this connection; among the emigrés, it is Fermi, Bethe and Wigner who were significant.

Session II. The Interaction of Theory and Experiment

Cyril S. Smith, Chairman

Smith: The question of the interaction of theory and experiment is one of the main headings on the [study] paper, and Professor Wheeler is specifically earmarked to talk about this.

Wheeler: I do not really think anybody can review the history of nuclear physics in this country without saying quite a bit about the contribution of Breit, as Viki [Weisskopf] has also mentioned. Merle Tuve has also mentioned his influence in connection with the accelerator work. Of course, I remember very much his calculations going on continually on penetration of protons into atomic nuclei to estimate reaction rates, trying first this kind of potential, then that kind of potential, to get a clear idea of which was the more reasonable potential to describe the interaction of the proton with the nucleus from the experiments. A feature of his was a driving conscience concerned to get things straight. He kept going over and over and over again, whether it was a hyperfine structure of lithium to get the magnetic moment of lithium better or whether it was to get the potential barrier of a lithium nucleus for quite a different thing—namely, the penetration of the proton—or whether it was quantum electrodynamics. He kept his finger in all these things at the time.

Of course, one of the most striking things, I think, is just to look at nuclear physics as it stood at the particular time in history, 1933 to '34. I particularly remember Breit with this wide range of concerns, and I remember especially the model of nuclear physics which was in one's mind at that time. To be sure, Heisenberg had come out with a picture of neutrons and protons constituting the nucleus the year before. Yet somehow this had not been built into any workable model for calculating nuclear reactions. The only handle that one really had at that time was this penetration factor. One can now talk of a Hamiltonian. Well, one did not have a Hamiltonian really to deal with except this model of a potential barrier and a single particle in it. Whether it was an alpha particle in a potential barrier or the proton in the potential barrier, there was no picture of the interaction between the particles inside of the nucleus. This was what was lacking.

But there was of course the very great impression that was made by the resonances at the time—the work of Bothe, Gentner and some of the people at the Cavendish—at which energies one got penetration of alpha particles into the nucleus and at which energies one got proton groups coming out of the nucleus. This was all one grand mystery, how to fit these together into any reasonable pattern, and a deeply puzzling side of the story.

I did want also in this connection, since I happened in 1934 to be part of the time with Breit and the other part of the time in the Copenhagen group, to mention the very great impression it made at Copenhagen when Møller brought back from Rome the story about what Fermi had been finding on the slow neutron resonances and how this was absolutely upsetting and how Bohr paced up and down and tried to find some way out. And although it is true that the Breit-Wigner formula came in the literature earlier than the compound nucleus model, I am not sure that the Breit-Wigner formula in fact came sooner than the compound nucleus in terms of the moment it was created. It took quite some time between the inception of the idea and the actual writing up.

Segrè: I know that we knew the Breit-Wigner formula and we did not pay much attention to the compound nucleus.

Telegdi: I must add that the derivation of the Breit-Wigner formula as given was not believable either.

Segrè: Oh, the derivation, nobody ever followed the derivation. Fermi had his personal derivation with second order perturbation theory.

Weisskopf: I believe that you are right, but I would like to check it because I know that the compound nucleus idea was discussed in Copenhagen when I arrived there early in '36, end of '35.

Segrè: You see, the resonances were found around this time.

Wheeler: Møller brought back the story of the resonances from Rome early in 1935.

Weisskopf: And there the compound nucleus theory was told.

Segrè: The slow neutrons were found in October, '34, and then the selective absorption early in '35.

Morrison: The capture of slow neutrons—Breit-Wigner— is dated 1936.

Wheeler: And I think the compound nucleus picture was agreed on before then.

Weisskopf: You know how long it took Bohr to write a paper.

Wheeler: Exactly.

Weiner: It might be of interest to think a bit about the effects of experimental results on models of the nucleus. Perhaps we can look at the case of resonances in that period and the effect it had on Bohr's thinking, to see if this represents a general pattern in the development of a model; and then, in turn, how much does the model, once it is created, influence the types of experiments that are done? Do experimentalists feel that it is their role to test the model? A number of questions come in.

Wheeler: I am very glad you bring this up because somehow we have got all our talk and discussion—and I think appropriately so for many purposes—focused around this 1932, '33, '35 period. However, I must say that if one got in a group of younger people in this field and had them around this table and asked them for the history of nuclear physics—for instance, take these six younger nuclear physicists—Brown, Kerman, Bromley, French, Rasmussen and Stelson—I think it would be an absolutely different history we would be hearing. There have been quite different developments, and I think one perhaps is getting an absolutely misleading impression about nuclear physics being a dead field from the way we are discussing it this morning; whereas, in my impression, nine-tenths of all one is ever going to find out about nuclear physics is yet to be found out.

Be that as it may, to return to the experiments on neutron capture: there were those endeavors to find out in Copenhagen just a very simple way to estimate what the probability was of a proton, or a neutron rather, going into a nucleus, the chances of radiating. The very clearest indication from those calculations is that you just get an absolutely fantastically lower order of magnitude of capture cross section than what one observes. And then this leads to this idea that the mean free path of particles in the nucleus, instead of being long as it had previously been assumed, was very short.

And one remembers the reaction of one's colleagues to this model, many of whom did not like it a bit. Especially I see Viki Weisskopf smiling. He knows that Eugene Wigner was one of those who at least liked it because Wigner was working so hard at understanding the regularities in the magnetic moments of nuclei, and these regularities were just incompatible with any such picture of a liquid drop and a particle with a short mean free path.

And so one really had for quite a period of time these two absolutely different currents of nuclear physics: when one group of people followed one point of view and another group followed the other point of view, the independent particle model versus the compound nucleus model. It was almost a failure of communication—well, not so much communication, neither could convince the other. And it was very healthy that it was this way.

Kevles: Is this the same point that Professor Weisskopf made before as to the divergence between the fundamental particle people and the phenomenologists?

Weisskopf: Not quite. It has something to do with it, but it is not really that.

Telegdi: The compound nucleus is more phenomenological.

Weisskopf: The compound nucleus is a shade more phenomenological, but it is really not. At that period, both groups which Johnny [Wheeler] mentions were essentially working on a phenomenological basis. And what I meant was those groups versus the people who were thinking of nuclear forces explained by mesons, meson exchange and this kind of thing.

Telegdi: If you want [to hear from] someone who is a little younger, I can tell you [about] the people who started to study after the war, like myself. At that time it had already become clear that the great thing right after the war for people like myself who studied then was the discovery of the pions and of the artificial pion, the beginning of a new kind of nuclear physics of elementary particles, the merging of cosmic ray research which was off in a corner, with elementary particles made by machines. The post-war atmosphere had this innovation you know, as we just have heard about the Cockcroft-Walton. There the great thing was to analyze cosmic rays, since you could then do this on earth. The most exciting thing that we heard about was the Conversi-Piccioni experiment. That was the thing.

The other thing I would like to say, which is perhaps more pertinent to the historian: I once asked Julio Racah in 1950 or so why he was working on the spectrum of iron, a very esoteric thing, instead of working on nuclei. And he said to me, "You see, when I was young in 1937, Wigner and Hund wrote two fundamental papers in nuclear physics which are full of group theory, and I liked that and I studied those papers and learned the methods. But then I wanted to apply them to nuclear physics, and I found that nothing was known about nuclei. So I started applying those methods to atomic spectra where there was some evidence and I stuck with it." He had all this grandiose machinery but no nuclear levels were known. So this may be very characteristic.

Wheeler: I must say I remember so much. There was the great white hope in the air that one was going to find nuclear forces very soon. The proton-proton scattering experiments were going to give one this answer. And that was really very much in the air. The theoretical people held it as an ideal. The experimental people held it as an ideal—to find this out. And the theory of scattering was a center of interest because out of the theory of scattering one could learn this law of force. And I am sure that each of us remembers his own approach. I remember particularly the experiments that had been done by the Cambridge people on the scattering of protons by alpha particles and alpha particles by alpha particles, where one could dope out these things to learn a great deal about the interaction.

But it was this dream of a simple story of the interaction that really got shattered by the compound nucleus model, for one group of people, but which was still kept very much alive by the other group of people thinking of the shell model, the independent particle picture. And those two currents stayed alive.

I remember in 1939 seeing how alive that division still was when it came to the study of fission. Would the different resonances of the uranium isotopes show different fission characteristics or would they all show the same fission characteristics? According to the one picture—the compound nucleus picture or rather the liquid-drop picture—one resonance and another resonance and another resonance would all be alike. There should not be any great difference. And according to the other colleagues who looked at the individual particle picture, one should be as wildly different from another as you could have it. They were just different spectroscopic levels.

Well, of course, one did not know for quite some time how the story really stood experimentally. So the dichotomy persisted.

One of the things of very lively interest was trying to study these collective aspects that Viki mentioned—the evaporation model. And there was quite a strong interaction, I think, here with the experimentalists because the picture of a particle going into the nucleus and heating it up, and then a particle coming out with this, that, or the other energy governed by an evaporation model and Maxwell distribution, was so simple and appealing to many of our experimental colleagues that it was something that one could latch hold of and do something about, make measurements on. And so measurements were made all over the place on this sort of thing. So this was quite a guide to the work that was going on.

On the other hand, I do feel that the Van de Graaff outfits with the precision determination of proton energies gave one quite a different approach. There, one was looking at one single resonance or another single resonance or another single resonance, and this just did not lend itself to the gross statistical treatment that one had for the compound nucleus picture.

So this kept alive a very strong current on the experimental side to interact with the theoretical groups that were interested in the individual particle picture.

Kevles: On this relationship of theory and experiment, speaking historically, one can show that for many decades in the United States, before, let us say, the 1920s, there was of course a very strong tradition of experimental physics in the U.S. and almost no tradition whatsoever of theoretical physics in the sense of mathematically created theoretical physics.

Segrè: Except Gibbs.

Kevles: Except Gibbs, yes, who was unappreciated and unrecognized by his own generation. And in experimental physics one could take the diffraction grating, for example, and just do measurements of spectra on the essentially Baconian assumption, which I think is very strong in the physics community here, that sooner or later, if you collected enough data, the answers would come out. The theory would emerge. What seems to me to be different about the development of experimental physics and nuclear physics here is that sociologically and economically speaking the equipment just got too expensive for everybody to have his own accelerator, apart from small accelerators. And what you had was the introduction of a kind of authoritative structure into the experimental community in the United States. It was the important and recognized physicist who really determined how you were going to use these machines and who insisted that there be meaningful questions asked. Is this a reasonable hypothesis?

Smith: I have never seen so many heads shake [negatively] .

Kevles: Certainly since there were a limited number of accelerators, and still are, you just could not go in and take a machine and do whatever you pleased.

Livingston: It is not true. There are hundreds of accelerators in this country.

Birkhoff: I'm not a historian or a physicist, but I would like to urge that history be considered an experimental not a theoretical science, especially in this connection: you look at the facts and not try to generate theories about them. I think it's very dangerous.

Kevles: I'm trying to look at the facts.

Tuve: I do not believe what you call an authoritative structure really existed. From the nature of the activity itself, it was a necessity for people to work with each other. So somebody set up a program, but it was not governed by...

Kevles: I do not mean authoritative in a dictatorial sense.

Tuve: The theory, in my opinion, has never been true in the United States—if you speak of, for example, the German tradition of the *Geheimrat* which determined everything in the laboratory and then locked it and went to lunch. I do not believe that even the existence of unique accelerators led to such a thing here.

Wilson: I would disagree. I think that there was a clear separation before the war in this regard and after the war.

Tuve: I'm speaking of pre-war.

Wilson: Yes, Then after the war, for example, with the construction of a place such as Brookhaven, where I think there was a tendency for laissez-faire to be invoked for the experimenters in the early days, there was a complete mess made in the use of the cosmotron. Eventually a committee, the worst of all possible things, was appointed. The committee, particularly under Rod Cool, worked because of a collaboration between theorists and experimentalists and the committee system essentially took over from that point to decide what experiments would be done. I think that that was imposing a kind of authority upon the physics, and I think that that has become tighter and tighter as time has gone by. I think there is this [aspect] ten years after the war.

Segrè: After all, an accelerator costs more than $200,000.

Livingston: That is just high-energy machines.

Smith: Not low-energy.

Segrè: Not yet.

Wilson: May I just say one more word? You have to look at the large international meetings that came about. Before the war there were occasionally such meetings, but they were not so widely attended, particularly I think, by many experimentalists. Right after the war there were a few theoretical meetings—at Shelter Island and, for example, one or two experimentalists might have been there. Then a few years later the Rochester meetings began, at which there were about an equal number of theorists and experimentalists. The two groups began communicating. A physicist reported on his experiments there, and then learned what experiments to do next. There is a difference between the experimentalist before the war, who could think up his own theory or use his own intuition and the kind of experimentalists that developed after the war, few of whom have made contributions if they ignored the theorists.

Kevles: Then the corollary of what you are saying is: before the war this very close collaboration between theorists and experimentalists, that we are assuming here, was not really true.

Wilson: It was very rare and fairly accidental.

Weiner: And not as necessary.

Weisskopf: Of course that is an interesting point but is it really so? After all, since there were far fewer people, it is hard to generalize. But, as I remember the situation at Cornell before the war, there certainly could not have been a closer collaboration between Bethe and his theories and the experimentalists.

Telegdi: You said it was specifically Bethe.

Wilson: Let me ask Tuve first. You did one of the first and fundamental experiments on the determination of the nuclear force and the scattering of protons on protons. Now, did you do that because you were just thinking that that was the simplest and most basic experiment you could do? Did that come from you yourself or did that come because Gregory Breit said, "See here, do this experiment. It's the most important thing."

Tuve: Well, I'm sorry. That goes back to my own great pleasure in reading about Rutherford's alpha particle scattering. So when I had to make the decision of whether to go to Cambridge or not, what I wanted to come to Carnegie for was the billiard ball scattering of protons; and I was surprised to find that eight years later it still had not really been done in the lab. Herb came to our lab. We were troubled with the voltage scale because to do these experiments you really had to have a decent voltage scale. So Herb came to our lab, I think in the summer of '35, in order to set up a wire-wound voltmeter which would calibrate the resonances we had found in fluorine and lithium so that we could use it for further proton scattering.

Morrison: Why did you need the good voltage scale?

Tuve: Because if you calculate the classical scattering values, you have to know rather precisely what voltage you have. I am just speaking now of the origins. This was an important idea in physics for many years.

Telegdi: If you don't find anything, it will be clear that the energy was too low. That is why you need it. You don't see any anomaly in the scattering.

Tuve: We did not know whether we would have anomalies. This is why we had trouble.

Weisskopf: Can you tell us something about the interaction with Breit?

Tuve: I came to Carnegie to work under Breit in atomic physics, as it was called. Breit was my chief, although Fleming, the Assistant Director, was the one who had persuaded me not to go to the Cavendish but to come here because, if I couldn't do it in one year, here I could keep on longer than one year. The purpose was to put a million volts on a vacuum tube in order primarily to do the billiard ball scattering with protons, copying the Rutherford classical alpha particle experiment. Breit was a partner in all of this. He made the calculations. This was some years later, after we got the voltages on tubes. But he made all the original calculations of phases and phase shifts.

Segrè: Did you know that you could expect anomalies because they were identical particles and you would have a Mott scattering and not a Rutherford?

Tuve: Breit did, yes.

Segrè: I was asking if at that time at Carnegie Institution you thought of that.

Tuve: Yes, but that is still a classical calculation.

Segrè: No, it is the identity of the particles. It is the Mott scattering.

Telegdi: Historically it is the other way around. The antisymmetrization of protons was done before the Mott scattering.

Segrè: And then it was verified by Chadwick for bosons in 1930.

Wilson: Gerthsen in Germany did the experiment of proton on proton at 30 kilovolts in 1931.

Weisskopf: Still there is a point here that worries me: the interaction between theory and experiment in the pre-war nuclear period. Some people say there is no interaction and others that the interaction is quite essential. I belong to the second group.

Tuve: Aren't you making an unnecessary distinction? Actually, ideas stimulate both experimentalists and theorists, and it was quite clear that the simple notion of scattering is an idea. It is neither theory nor experiment. Both move ahead together.

Wilson: No, we are talking about this imposition of authority, which is how this discussion came about. The point that I was trying to make was that there is something in this idea, but it did occur much later after, say, 1955. I think before the war, many experimentalists read the literature but decided what they wanted to do on the basis of their own understanding and did not come to a theorist and ask, "What should I do now?" For many of them this would just have been an impossible approach to physics.

Weisskopf: Absolutely. The point is also not that the theorist felt this. What I want to emphasize here is that it is a collaborative effort. The main point which I, at least, found extremely new and attractive in the '30s when I came to this country was, for example, the living together and the close collaboration of the theorist and experimentalist. It is true that the experimenters, the good ones, came to their own decisions about what they should do, but this was after long discussions; and the same thing is true about the theorists, about what they should work on. This came out of a very close collaboration.

Telegdi: The fact of the matter is that when experiments are reasonably cheap, the experimentalist could decide to play around and collect data by what we referred to as the Baconian method. But when experiments go beyond a certain point, one has to justify the importance, and that is usually done by dragging in theorists.

Herb: Would not what you say, though, apply mainly in high-energy physics?

Wilson: Probably so. Not in low-energy physics.

Herb: These are great monsters of machines.

Segrè: Mr. Telegdi, I would amend your statement to dragging in committees, not theorists.

Telegdi: That is still a very important sociological point. The point is that a scientist cannot say that he wants to do something because he finds it interesting, but he needs others, plurally or singularly, in committees, to support him in the desirability of the experiments. If physics is cheap he can do it.

Segrè: Because he has to spend so much money.

Telegdi: That is what I started to say. The only point where we differ is that I say dragging in theorists and you say committees. When the money gets even bigger, then there are committees of theorists and experimentalists that have to support it.

Segrè: You can see this at Berkeley. If you want to do something with the cyclotron, you don't ask anybody. If you want to do it with the bevatron where we know it costs so much more, you have to get permission.

Holton: There is also a problem here owing to the rapid rise of theory at a time when the proper type of activity is still said to be experimental. There is now a man still working as a professor of physics who was the first man in America to write a thesis as an acknowledged theoretical student. That is Van Vleck. His thesis, and it may be off by a year, is said to be the first one which was out-and-out theoretical. He took it under Kemble, who was just a little older, but who, in turn, although he was very interested in the theory of specific heats, took his Ph.D. work supposedly on an experimental problem. So I think that this sudden rise of people who had gone through theory is attended by their background in experimental courses. If you look at the course catalogue, for example, you find that those people who come out with a theoretical thesis have gone through quite a lot of experimental work just by the graduate requirements. With experimentalists, they are not from the beginning typed as experimentalists. In their own graduate work they have done quite a lot of theory. And I think that when you look at this period of 30 years, where one switches over from a predominantly experimental to a much more mixed state, this factor may be worth looking at.

Weiner: I would like to offer some statistics, or impressions anyway, on that. Of the dozen people interviewed so far, a good many of the Americans had taken their undergraduate degress in engineering, and this included theoreticians as well as experimentalists. And if they had not taken their undergraduate degrees in engineering, they had certainly started in engineering and only later switched to physics.

Telegdi: In Europe, too: Dirac, Wigner, Szilard, Teller.

Kevles: Would it help to clarify the discussion a bit, I wonder, if we say that perhaps we are talking about theory in a very special sense of theory? That is, theory in the sense of being creative in complicated mathematical expressions of physical ideas as opposed to what Dr. Tuve said previously, and that is an idea of nature or a model of nature. For example, when Rutherford did his famous scattering experiment in 1911, he came up with the theory of atomic structure, you might say, with an idea of what the atom looked like. And I think to a great extent people operated all the way through, perhaps into the 1930's, with a mixture. I mean the physics community operated with a mixture of mathematical theory and models and ideas of what the nucleus was like. And what has happened in more recent years is that the special sense of theory has taken hold more and more.

Kuhn: I think you are quite right in wanting to distinguish when the question is asked, "What is the role of theory?" between types of models, which are available and accessible both to the experimentalist and the theoretician and which may be drastically transformed in a few years, and the notion of high-powered theory. On the other hand, I think historically your description of the developments, making the whole thing come this late, is probably wrong. That is, you get, at an earlier point, Heisenberg's interest in fundamental problems. You get a very high-powered quantum mechanical theory in the late teens and through the '20s, which is something that very few of the experimentalists can do. Thus you get briefly these same problems of interaction and the same sort of high-powered theory in Europe through the '20s with respect to the problems of quantum mechanics.

Smith: I would be inclined to the view that the important stage of theory is the theory in the mind of the experimentalist. I was very much impressed by what Bob Wilson said this morning about the kind of model that he had in his mind, which was related to what we call really good theory. It's this kind of conceptual, almost sensual, feeling about what goes on that determines the next stage in the lab.

Telegdi: No. It is a matter of tradition. I was told by my major professor that when he got his degree, which must have been around 1912, 1913, the difference between a theorist and experimentalist was whether he did or did not understand Maxwell's equations. Now, when I studied, the difference was roughly whether you did or did not understand Dirac's equation. Most of our experimental students today can manipulate the Dirac equation. I would not say that they can understand it, because that is a big word; but they can manipulate it, and who knows what comes next? I would say what one calls an acceptable way to think about things is a matter of habit. Concepts get accepted, and things which must have been a tremendous mental effort 20 years ago are now household ideas for graduate students. So I think there is no absolute definition. It is a matter of age really as to what you find acceptable. I told Weisskopf many times that he wants to express everything in certain simple terms. Those are the terms that were simple and created in 1931, at which time they were a great intellectual effort.

Wheeler: Those seminars that we used to have in the old days when physics was not so fragmented and everyone was in it together and ideas were kicked around and one tried to state the idea in simple terms: somehow we do not have it quite the same way today. I guess we all agree that things have changed. Theoreticians have theoretical seminars and experimentalists experimental seminars. We have lost a great deal in that.

But before I turn away from one side of this subject of theory and experiment, I think that Rutherford deserves some credit for one of the key ideas of quantum theory. In the 1905 theory of radioactive decay he insisted that the concept of the radioactive decay constant had to be accepted. You could not understand it; you could hardly believe it. It was so incredible. It did not matter how old an atom or nucleus was. They did not even know about the nucleus then. But however old it was, there was the same decay constant. And just simply to insist that the experiments told you this and you had to take it was really the key idea of transition probabilities in the quantum theory.

Morrison: It is purely a statistical idea rather than a quantum idea. It is interesting that Kelvin never believed it. He was a very old man, of course. He attacked it bitterly. His last years were concerned with attacking statistical mechanics.

Wheeler: But of course Kelvin in 1903 attacked Maxwell's equations. He said you could not believe them either.

Segrè: If I remember right the purely statistical feature of radioactive decay was explained with an incorrect model proposed by Lindemann in 1915. It is quoted in Rutherford.[24] He said, "After all, you don't have to make it absolutely probabilistic. If you have many independent causes, they will explain the exponential decay."

Morrison: That in fact was Poincaré's theory of probability.

Wheeler: Poincaré has to get credit also for the discovery of radioactivity, because that is Becquerel's story. There was a meeting in the French Academy where Poincaré was discussing with his neighbor Becquerel the amazing radiations just discovered by Roentgen and then he said to Becquerel, "Look, you've been studying this business of delayed phosphorescence. Why don't you try to see if you can't tie up this effect with Roentgen's work?" (This is not an exact quotation.) And so Becquerel did the experiments which led to the discovery of radioactivity. This was told by the son of Becquerel to Gorodetzky; and that is where I got it.

Caws: I would like to raise the question in the inverse sense. We have been talking about how the experimentalist can or cannot decide on his own what to do, especially in light of the cost of the apparatus and so on. But how about the theoretician deciding what or what not to do when his own structure has become large and complicated—as in fact they do? Which phenomenon then does one choose to try to build a theoretical structure of a very complex kind, involving matrices or something, that is no longer a simple model of nature but something pretty large and mathematical which takes a lot of investment of intellect, just as the other thing takes investment of funds?

Smith: Intellect is cheap. (laughter)

Segrè: As an experimentalist, I would say that it is very good to listen to theorists, but you cannot be guided by them because they have two habits. If at a certain moment they get the idea that this is the central problem of physics, they see nothing else. For instance, at a certain moment you had to see the p wave in proton-proton scattering. That was it, you see. Secondly,

24. Rutherford, Chadwick, Ellis, *Radiations from Radioactive Substances,* 325 (Cambridge, England: The University Press, 1930).

they do not realize (not all the theorists, because there are various kinds of theorists) that certain things are possible with a certain degree of accuracy, and certain others are not. So that any experimentalist who is worth the name of physicist should decide by himself. He should learn enough theory to know what is valuable but should not be guided by the theorist.

Telegdi: He should be guided by the ideas of the theorist but not by the theorist himself.

Herb: He could be guided but not controlled.

Wheeler: Could I bring up one other side here about this later division that we'll perhaps get to further on, about elementary particle physics versus nuclear physics. I was referring a little bit earlier to fission, early work on fission in 1939. Of course, it did not come into the public domain until after the war, but the interesting thing is how long it took to come back into the public domain. It was not until 1955 that the Atoms for Peace conference came off and fission could be discussed publicly. So here was a broad part of the whole domain of nuclear physics that was just cut off, one of the most interesting parts of nuclear physics from the point of view of discriminating between the shell model picture and the liquid-drop picture. It was foolish, absolutely foolish, for anybody in that whole ten-year period to undertake work on that field in a university laboratory, because who knew? How could he tell he would not be duplicating something that was being done in a classified laboratory? So people were pushed out of that part of physics willy-nilly, in effect, and I think this contributed a little bit to people going into elementary particle physics—not that there wasn't plenty of other motivation too.

Weisskopf: There was plenty of other nuclear physics.

Wheeler: There was lots of other nuclear physics, of course on reaction rates of iron and what not which was going on, but one of the most interesting parts was just cut off. It is very interesting, as a matter of fact, even today how the role of fission physics in getting support for all physics receives as little attention as the role of sex does in bringing about our existence.

Telegdi: We can discuss that. (laughter)

Smith: It seems to me we have not answered the question that was raised about how the theoreticians select the area they're going to work on.

Holton: More and more they do it by asking for a problem.

Weisskopf: Yes, I feel it is not the investment. It is simply that it is very difficult to be inventive in theory. Some people say it is more difficult now than before. I don't think so. I think it was always difficult. And therefore many of the theorists are following a fashion or following a leader. But this is not because if they wanted to, they couldn't do otherwise. It is just because they are not inventive enough. I do not think this is in any respect an authoritarian regime except a voluntary authoritarian regime.

Telegdi: An aristocracy.

Weisskopf: An aristocracy, yes. It is just very, very hard to be original in theoretical physics.

Telegdi: You think it is easier in experimental physics?

Weisskopf: No, I am just saying that it is hard in theoretical physics.

Wheeler: I think it would be very interesting in this connection to ask how many people worked in nuclear physics by preference, or whether it was really a way to make some little progress because it was easy enough theoretically to make some progress in nuclear physics, while always keeping burning on another burner some other part of theoretical physics. For example, I could ask Viki: Would you say that the nuclear physics that you did, for example, in the postwar period in the first ten years from '45 to '55, was because that was your great white hope for the future or because you did not at the moment see a way to make bigger progress in another area that you would really have liked much more to be working on?

Weisskopf: That is one of those things one cannot answer in any direct way, as it is very personal. I think I did it because it was successful. Something came out of it. In fact, I personally always had a bad conscience in this period. I found this nuclear structure work successful, and I started something and I had to continue; whereas at the same time elementary particle physics was becoming very interesting and I had no time really to delve into it. I had a certain split personality.

Telegdi: Maybe if you had at that time been 15 or 20 years younger, you would not have continued the nuclear physics but would have gone into elementary particles. It may have been a matter of age.

Weisskopf: That is probable, but I don't know.

Smith: The physicists have rightly had most of the floor this morning. We will be adjourning for lunch before very long and I think we should hear more, particularly from the sociologists, who have been quite quiet this morning.

Auerbach: I have a sociologist's question though I am not a sociologist. We've been talking about the interactions between theory and experiment, the experimentalist talking to theoreticians. I have also heard it said that during the '30s nobody read journals.

All: No, no.

Wilson: On the contrary.

Telegdi: There were only two journals.

Tuve: You have to distinguish, though. In the late '30s the German literature really went to a very low level.

Herb: The French also.

Tuve: So there is a partial truth to what you've heard. After the Nazis took over, it became increasingly less interesting to read the *Zeitschrift für Physik.* By '38 and '39 it really was entirely uninteresting.

Wilson: But there was a time that you're talking about when one could read the whole *Physical Review* and generally did. And not only that, you read lots of other journals all the way through.

Tuve: There were lots of journals. We were primarily dependent on journal clubs.

Auerbach: It has occurred to me that perhaps there was a transition from the great importance of journals in the '30s to perhaps the conference or the very small specialized seminar. And I am just curious as to when this took place.

Livingston: When air travel developed, we began to communicate.

Wilson: Also, when the journals stopped being printed. There was a much greater lag—I think maybe a four-month or maybe a six-month lag—between the time that you submitted a paper and when it came out. And because of conferences and because of preprints, those two things, one could get informed very much faster by different techniques.

Auerbach: Did journals actually become slower as methods of communication or did everything else just become faster?

Wilson: I believe that journals became slower after the war.

Segrè: They started the *Physical Review Letters.*

Wilson: That was quite late, though.

Weisskopf: I think they never became slower. Only our tempo became faster.

Wheeler: Does anybody remember making a long distance call before World War II?

Weisskopf: Were there long distance calls?

Holton: I remember at the time (1942-45) it was rather astonishing that everything that goes to Santa Fe had to have an air mail stamp.

Telegdi: I would like to raise a couple of questions which interest me but which I cannot possibly answer as they are sociological type questions. We talk about this role of the European immigrants, the big wave of refugees to America and what impact it had or did not have and what is to be attributed to them. It does not seem quite obvious because very mature people were brought here rather than green people. But what has always puzzled me is that a fair number of these people who came to the United States as a first step went to France, where on the whole they were not accepted and not settled.

Weisskopf: England, also.

Telegdi: Well, I am not as familiar with England perhaps, but I think that people settled in England a little easier than in France. In France the local establishment was really not very favorable to the settling of these people, and many of them I think personally would have preferred to stay in Europe until it was quite clear we should leave anyhow, which you had to do because of the war. And French physics, which in 1932 was actually at a very high level, as we all know, got nowhere because this wave of immigrants was not received. We do not even have to talk about Europe and America, [we can] just talk about one European country that did not absorb this wave of people. It is a perfectly possible situation.

Tuve: There are other reasons why the French went downhill besides their failure to accept outsiders.

Telegdi: Well, maybe one could have offset the other.

Weisskopf: The question is: what is the cause and what is the effect?

Telegdi: Then it is not a sociological question, but it is perhaps worth entertaining as a sideline.

Weiner: Part of the question too, I think, is what effect did the interim period have on the physicists who eventually came here [for example] Bethe, Weisskopf. A number of people stopped either in various laboratories in England or in Copenhagen and then moved on, because of the fact that there just were no jobs available. Despite Rutherford's intense efforts with the Academic Assistance Council, there just were not as many opportunities. People served an apprenticeship under Blackett, under Rutherford, under Bohr, under other people, and then came to this country. And perhaps we ought to examine this interim period as a very important training period at a very important time in the life of a person when he has no roots any more, when he is really in a state of escape. What kind of a learning experience did he absorb during this stage of his life? And then consider what his impact was when he got to this country. And perhaps this is done through individual autobiographies more than anything else.

Holton: There is a point about Mr. Wheeler's remarks which I think has some sociological interest, and that is the following: He spoke about the two camps in nuclear physics which worked parallel but which did not believe very much in each other's model. Now of course one can find a great deal in the literature, and that will always be the case; but there is a lot that does not get into the literature about the confrontation between these camps. Were there memorable occasions when at conferences it did come to exchanges which are not in the literature? Are there documentations of how there were converts made from one camp to the other? Or what, in fact, was finally responsible for the dissolving of these confrontations? I think that kind of unwritten thing will not be dug up by the archeologist.

Telegdi: The history of rejected papers might be very important. I personally know that the American paper that proposed the compound nucleus was rejected by *Physical Review.*

Weiner: There's a pilot study under way now to take a look at the rejected papers, preserving the anonymity of the refereeing process. It's only for a recent period, a five-year period, in the '50s.

Telegdi: But if you could establish that some of the major good ideas have actually been rejected, it would be a very healthy influence on our journals.

Wheeler: It might not be.

Segrè: Another example of rejection: The beta ray theory of Fermi was rejected.

Morrison: By whom?

Telegdi: By *Nature.* There was a great controversy.

Segrè: I can give you another nice example. A big paper of Wick's on the exchange forces in proton-scattering with the angular dependence was rejected both by *Nature* and by *Naturwissenschaften* and then was published only through the personal intervention of Heisenberg.

Morrison: This raises an interesting question, too. I would like to find out the date at which the majority of American physicists took American Ph.D.'s. It is some time in this period. People older than I am did not.

Weiner: It happened in the 1920s.

Kevles: No. In fact, you can date it earlier, I think, probably before the war [World War I] so far as taking the degree is concerned.

Mendelsohn: If you should turn that around and say the date when major American physicists who made contributions took degrees in the United States, the dates will be different.

Kevles: They all went to Europe afterwards to study, but the Ph.D. degree, taken in America, started to be known after the turn of the century.

Caws: There is another sociological question which may come up this afternoon and which Mr. Wheeler made it possible to comment on. That is the whole question of secrecy. I do not know whether it is appropriate in this discussion or whether it has been worked over too much in other contexts.

Weisskopf: No, it should be discussed, though I think the fact that fission was not taken up was only partly due to secrecy. I think there was also a certain feeling, rightly or wrongly, in the community of nuclear physicists that it was not simple enough and not much would come out. Now, of course, you could say that this was because people did not know enough about it and therefore it could be traced to secrecy. That could be, but I am not so sure that secrecy has played such a role in keeping that field back.

Wilson: Also, I think there would again have been a tendency, if you were starting work in fission at some laboratory, to have to start at a low level compared to people that you knew were doing that kind of work in the national laboratories at a fairly high level. You would have to bring yourself all the way up to that level before you could compete; whereas you could start on another problem and compete immediately.

Auerbach: How do you compete if it is not being published? What sort of competition is it?

Wilson: I think the people likely to have been working in the field of fission in physics would have known, or would have known someone who could have told them. After all, there was the Smyth report, and there was an awful lot of information out about what kind of work was going on in the national laboratories. I think they would have had enough information to make a judgment that this was not a good field for them to work in compared to another—for example, measuring of resonances—to do the kind of work in great detail that was done, say, at Wisconsin. That work was probably much more productive and probably did contribute more to the field than the work in fission that was done subsequent to the war.

Weiner: Do you think the availability or lack of research funds for fission work in a university, not in a government laboratory, played a role in determining whether one tackled that subject or not?

Wilson: I just doubt that there were many people who made applications. I doubt that you would have found any applications for work in fission. A man for various reasons would not apply, partly because he would not want to do work with uranium, which would, after all, probably involve any university laboratory in a very serious problem: the possibility of doing classified work in research. But I think it was more than that. They would have found more interesting work to do.

Herb: Yes.

Wheeler: You take the Big Mike shot of the first of November 1952, (the first thermonuclear explosion). How little thinking there was going on about superheavy nuclei and fission is shown by the fact that nobody was prepared to look for the new nuclei that were going to be created there. It was only after the event that one realized it and then could get going on the debris. But if one had really been prepared, he could have gotten so much more out of that shot.

Smith: It is getting on toward time for lunch. But before we adjourn, Professor Segrè has a quotation that he wants to read.

Segrè: It is not a quotation. There is an article in Italian written in 1929, which has this great thing of being dated. It describes exactly the way experimental physics appeared to Corbino, and that means to a certain extent also to Fermi, and why. It's entitled "The New Assignments, the New Purposes of Experimental Physics." It explains the transition from atomic to nuclear physics—in '29, before the big discoveries—very clearly and very well. And I can quote you only a few sentences; it is a long article, very well written. Somebody should translate it.* It says essentially that atomic physics has been completed, and with the new conceptions of Heisenberg, Dirac and so on, one can assert that the main part of atomic physics has been found and only details are missing. Atomic physics has found its definitive form.

And in this formulation there is no place for new forces. Then there is a long explanation of why he says so: The only possibility of great new discoveries in physics is now in the possibility of modifying the nucleus. This will be the

*Translation was subsequently made by Fausta Segrè, University of California Santa Barbara, at the request of the AIP Center for History and Philosophy of Physics. It was published in *Minerva, 9,* 528-538 (October 1971).

true purpose of physics of the future. He explains a little about this and then says how one will do it. He says Rutherford did it with naturally accelerated particles, but this is not enough. He next discusses other ways, high temperatures, thermonuclear reactions and so on. He says they won't work. And then he says the only way which is open is to produce artificially accelerated projectiles, but in a much greater number and with greater velocity, which requires the making of discharge tubes with several million volts. There are only technical and financial difficulties for this, but those are not insuperable, and the objective of this great project is not only transmutation of chemical elements, but also the liberation of energy, which will come about from the recomposition of nuclei and from nuclear reactions. And then he keeps on discussing this, and says one can thus conclude that whereas there are no great prospects of tremendous progress in experimental physics in its ordinary domain—by this he means classical and atomic—very great possibilities are open in the attack on the atomic nucleus or in the study of complex molecules of biological interest: the two fields of physics of tomorrow. Then to participate in this general movement in the modern direction of physics, it is indispensable that the experimentalist have a fast and sure knowledge of the results which theoretical physics is acquiring and that they have bigger experimental means. To attempt to make experimental physics without knowing day-by-day the results of theoretical physics and without having large laboratories is the same as trying to win a modern battle without artillery. Then he goes on and explains this.

Wheeler: Was this before or after he brought Fermi to Rome?

Segrè: After. It is Corbino's writing, Corbino's style, but I am sure it was discussed with Fermi. And then he says—this is 1929[25] —"One doesn't know what will come out of nuclear physics, but anyway this is what has to be done." And then the final thing is that maybe the nucleus will not give us this great surprise; but if that is so, then the only thing to do will be to turn from physics to biology. And this should be done not by having biologists superimposing techniques, but by having in the same brain the knowledge of physics and biology. He explains this in detail.

25. O. M. Corbino, "I Nuovi Compiti della Fisica Sperimentale", *L'Energia Elletrica,* 998 (October 1929).

Weisskopf: This should be published in *Physics Today,* as soon as possible.

Segrè: Well, it is an extremely interesting article, and it explains the motivation of physicists. At the beginning there is also what you can do in physics. You can discover new phenomena; you can measure things; you can accumulate data which go into books and are of no use. And then he says physicists shouldn't play like big children making expensive toys.

[Adjournment for luncheon]

Refreshment time on the grounds of the House of the American Academy of Arts and Sciences during the May 1967 Conference. (Photos by Heka.)

Above: M. Stanley Livingston (left) and Merle A. Tuve.
Below: Philip A. Morrison (left) and Oscar Handlin.

Above: Robert K. Merton (left) and Cyril S. Smith. *Below:* Left to right, Erwin N. Hiebert, I. Bernard Cohen, John A. Wheeler, Robert R. Wilson, Valentine L. Telegdi.

Above: In the foreground, from left to right. Joan Bromberg, Philip A. Morrison, Emilio G. Segrè. *Below:* Raymond G. Herb (left) and Robert K. Merton.

Session III. Postwar Development of Nuclear Physics

Robert K. Merton, Chairman

Merton: Our morning session was largely devoted to the physicists in the room trying to reconstruct some of the developments in the prewar period as they actually occurred, and I suppose one of the interests in this oral exchange of remembered history was the frequency with which some of the same episodes were felt to have quite diverse implications. The transcript will surely show that a number of specific questions came out of this morning's observations.*

This afternoon's session is to be devoted to two quite discrete subjects so far as the prepared agenda indicates. The first, which will be introduced by Professor Wilson, will deal with some of the postwar developments of nuclear physics.

Wilson: If I'm going to start talking about postwar physics, I really ought to talk about wartime physics because the wartime physics had a very distinctive effect on the postwar period. The thing that occurs to me first is that a change of style occurred during the war. I have been making the point that an experimentalist before the war, myself particularly, felt that it was absolutely necessary to understand what it was that I was doing. I had to feel that I was generating the kind of experiments that I happened to do, and that I did those experiments quite independently, although with the full knowledge of what the theorists were doing.

To be personal, at the beginning of the war I was working on the scattering of protons by protons—one of those very simple conceptual experiments. Then, when I wanted to go into neutron work—I realized I had little idea what a neutron really was. It did not have a charge. What kind of a particle was it? I, as an experimentalist, had no conception really of the neutron. And so the first thing that I did at Princeton was to realize that this time I would have to rely on my theoretical friends. I could no longer enjoy the luxury of using my own concepts. I had to learn their concepts as fast as I could, but I also had to depend on them. I think one saw that dependence grow more and more. At Los Alamos there was a theory division. The theory division was of dominant

*Because of a malfunctioning of the recording apparatus, only part of Robert Merton's introductory remarks were recorded.

importance, in fact, in deciding the state of understanding of fission physics, of neutron physics, and what experiments would be done. The experimentalists fed in their own opinions, but I would say that the theory group—especially given the fact that the director was a theorist and that there was such a strong assembly of theorists—became dominant there. That is my opinion.

That situation maintained itself somewhat after the war. We have already gone over this a little. Just after the war, there were only theoretical meetings. Then, combinations of theory and experiment came to be presented at the Rochester meetings. In elementary particle physics, it was at these meetings that one reported what had been done, tended to digest the work, and then to go away from them with a pretty good idea of what had to be done in the coming year. I am exaggerating because, of course, that process goes on constantly in various laboratories, in personal discussions, and in one's own thinking. But I do think that the meetings became rather dominant landmarks in what had been done and what was going to be done in the future.

Now, as to the labs themselves: The wartime labs—the Radiation Laboratory and Los Alamos and the Metallurgical Laboratory—were not so important as one normally believes. For instance, I think the Lawrence Radiation Laboratory existed before the war and was not of an order of magnitude smaller than, say, the Metallurgical Laboratory.

There was a difference—and I think we drew a false lesson from that fact—in that in the large wartime laboratories one had a mixture of disciplines. There were the chemists and metallurgists, and we worked very closely with them. I believed, at least when I was at Los Alamos thinking about setting up a laboratory after the war, that one would set up a laboratory not vastly different from Los Alamos; that there would be chemists and probably biologists and metallurgists and that we would become a much more powerful group as a result of that. We were thinking about what we were likely to be doing after the war. I thought that physicists would be making reactors and doing interesting neutron physics; that they would be developing power sources and applying radioactive substances to treating cancer. I imagined laboratories in which all sorts of good things were going to happen.

When it came right down to it, though, that probably was the vision of the Fermi Institute in Chicago. I was invited to join that institution. When I looked at it, I was aghast at the possibility and decided to go off to the opposite extreme, which was Harvard. Bainbridge and I decided that we were tired old men and that all we wanted to do was go back and cultivate our garden—just be as pure nuclear physicists as we could. We wanted to take Bainbridge's cyclotron back to Harvard and just start off where we had left off before the war. I wanted to do proton-proton scattering, and he wanted to continue the problems that he had been doing just before the war. We put the whole business of Los Alamos behind us. We just turned our backs on it.

But we were not allowed to do that because, for example, they wanted to keep the cyclotron at Los Alamos. And then entered Ernest Lawrence, a tremendous enthusiast who convinced us that it was our duty to leave that miserable cyclotron at Los Alamos. We ought to build something bigger and better, five times bigger. He would help us do it; he would find the money some place. We didn't have to use the rather slim resources available at Harvard. And I think that is what Ernest Lawrence did in other cases, too. He sold the Rochester people an even larger cyclotron than ours, sold the Chicago Fermi Institute (Herb Anderson) a larger cyclotron, sold the people at Carnegie Tech and Columbia. And all those cyclotrons were essentially designed at Berkeley (not so much the Carnegie one, which came later). One was pushed very hard by Ernest Lawrence, who had a vision of the Radiation Laboratory becoming a very large, important laboratory and then spawning the same kinds of laboratories throughout the country. I think that without Ernest Lawrence, Bainbridge and I would have come back to Harvard, put in the little cyclotron and done rather small-time physics with it, probably very successfully.

Segrè: What do you think Fermi would have done?

Wilson: That's very interesting. I do not believe that at the Fermi Institute, which was built around Fermi, a large accelerator would have been designed and built. I don't know. But I don't think that the Fermi Institute would have taken the direction in physics that it did.

Segrè: I am not so sure. Fermi said, "I don't want to touch nuclear physics after Los Alamos. The future is in particle physics." He made another turnabout as he had done from atom to nucleus and this is something which is conscious, and one knows. He could conceivably have used real cosmic rays, but he certainly would have gone to particle physics.

Wilson: I agree that he would have gone to particle physics. That is one of the points I want to make too: Most of the people at Los Alamos, most of the generation of physicists before Los Alamos, and many of those who came to Los Alamos or the Metallurgical Laboratory, did go into particle physics. I think my point is that most of us were in particle physics from the beginning. Nuclear physics was not distinguishable from particle physics before the war.

Weisskopf: You have a contradiction there. Fermi would certainly have known that you would need a large machine for particle physics. So he would ask somebody to build him that large machine.

Morrison: McMillan had invented the large machine, and Bethe and Fermi decided to build one somehow, to get somebody to build one or something like that. So did Robert [Wilson], but he doesn't remember it.

Wilson: No, sir. I remember very well what I did. I felt suddenly very pure about the whole business. I wanted to get away from nuclear physics. I wanted to go back and scatter protons from protons. I felt that there was a fundamental problem there that had not been solved, and indeed it hadn't. I thought that if you started to build a big machine, it would have been something that was very interesting, something that would come along; but that that was not the direction of the pure nuclear physicist, which was particle physics.

Weisskopf: Fermi was more ambitious. He said, "I would let somebody else build a machine for me."

Segrè: Fermi's stand was very clear. He said: "Nuclear physics is finished. I'm going to do a few more experiments with neutrons because there is a strong neutron source. In the meantime, I'm going to learn the theory of elementary particles, whatever is known. I'm going to prepare myself for the theory. Somebody is going to build me a cyclotron." And this is what he did. And it was planned pretty clearly.

Telegdi: I think the idea of the Institute came up in '45 at Los Alamos, and I think the synchro-cyclotron was invented only in '46.

Wilson: McMillan[26] had invented it in '45 at Los Alamos and it was very much in the air. On the other hand, I don't believe that at that time the decision to build the cyclotron at the Fermi Institute had been taken by any means.

Segrè: This I don't know.

Wilson: When I was considering going there, there was no cyclotron.

Segrè: Fermi's decision was not to make the cyclotron but to learn the theory.

Wilson: But at the time at the Fermi Institute there was to be a betatron, which had been ordered from General Electric—about the worst decision that had ever been taken by anyone. If Fermi was responsible for that, God help his reputation. That is the fact, that they had decided to buy some horrible machine from General Electric, which never did work.

Segrè: It never worked. That's right.

Telegdi: I would not have been at Chicago if it hadn't worked. I did my work on it. (Laughter) That is not to say that it worked well.

Segrè: The idea, in this specific case, was more to learn the theory than to make the machine.

26. E. McMillan, "The Synchrotron—A Proposed High Energy Particle Accelerator," *Phys. Rev., 68,* 143 (September 1945); and Veksler, "A New Method of Acceleration of Relativistic Particles," *Journal of Physics,* USSR, *9,* 153 (March 1945).

Wilson: I agree. At Los Alamos, for example, I took a course in elementary particle physics from Fermi, given to three people, to bring us up-to-date with what had gone on before the war. Fermi was giving such a course there and also one in quantum electrodynamics that he had worked out before. He gave a refresher course in that subject, too. I agree with you that Fermi had decided to go into particle physics.

Telegdi: And also the machine in Chicago was clearly built with the view of producing mesons.

Wilson: But, I think an interesting point was the one that I made, and I wonder whether you would agree, Johnny [Wheeler], that most physicists, wartime physicists, went back to universities and decided to go into elementary particle physics. They didn't call it that.

Wheeler: Right. I can remember very much the last few weeks before I left where I was at Hanford thinking very much about methods of cheap acceleration of particles based on cosmic ray mechanisms. Could one get some absolutely new method? I also remember talking to people at G.E. about air core accelerators, and I finally concluded that the only way to get particles cheaply and without having to do a lot of work oneself was to use the cosmic rays. And so on the strength of that we got cosmic ray work going at Princeton.

Livingston: Bob [Wilson], there was another rather important development, which was the Brookhaven Laboratory development with Associated Universities which came right in at the end of the war. Rabi and others organized that, and their goal was specifically a laboratory with both sides: a reactor but also the biggest accelerator that could be built. Those were the instructions I got: find the highest high-energy machine that can be built. That was specific, and that was in '46.

Wilson: I believe that started a little bit as the Fermi Institute attempted to make a reactor. The idea was to make a reactor first and then, I believe, added to that was a large accelerator. I may be wrong about this.

Livingston: I have just been reading Ramsey's Brookhaven report, just out recently, called "The Origin of the Brookhaven Laboratory,"[27] in which he makes some points about the fact that, although they did want a reactor because they recognized that was the way to get government money, they also from the very start intended to have the biggest accelerator they could get.

Morrison: What was the date of Art Roberts' song? Because it is very clear in that.

Weiner: It is mentioned in Ramsey's article.

Livingston: I think I know what happened. !f I may give a little sidelight on the history here: I was asked by the group that was setting up the Brookhaven Laboratory to survey the field and report back what was the highest energy machine that could be built. I went to Berkeley and talked with them. They had their synchro-cyclotron just under design. They were aiming at 200 million [electron volts] at that time, and it was barely arranged in '46. That was the conversion of the 184-inch [cyclotron]. And I came back and reported: "We can build one of that style, 700 million volts." I was asked to give a talk at MIT. Roberts was still there, at that time, with the Robley Evans group. He heard the talk that I gave on how to build this monstrous big machine, a sketchy concept of it. It was within a few months of that time when he came out with the song: "Take Away Your Billion Dollars, Let's Be Physicists Again." I think I stimulated him into it.

Holton: What you are recounting is a beautiful example of leap-frogging. While one is just barely starting, the other one already takes the advantage of that design and goes into the next generation. Is that something that happened more than occasionally? Was that the way in which machines got started?

Livingston: Yes. Just one more brief word about the Brookhaven story to show how it got where it did. Even before I had the design reasonably well-started at Brookhaven for the 700 MeV synchro-[cyclotron], Rabi came out and needled Phil Morse, myself and others: "Take something that's a bigger

27. N. F. Ramsey, *Early History of Associated Universities and Brookhaven National Laboratory*, BNL *922*, (Brookhaven Lecture Series, March 1966).

challenge," he said, "Why don't you take this other idea—the proton synchrotron idea," which was part of McMillan's general theory of synchronous accelerators, "and make something big up in the billion volts [range]?" And he pushed us into it. I immediately started a second design group. Eventually that took over, and we ended up with the cosmotron as the first multi-Bev machine at Brookhaven.

Wilson: Did that come after the bevatron or before?

Livingston: No, our decision was actually taken just before the bevatron, but they were made simultaneously, and the [Atomic Energy] Commission acted on them simultaneously and funded both at the same time.

Kevles: Was it a uniform opinion among the physics community that you wanted to build these accelerators to study fundamental particles instead of nuclear structure? I do not know if one can separate the two fields.

Livingston: Well, it seemed obvious to us that what we wanted to do was to get a new higher range of energies and here McMillan had given us a wonderful opportunity. We could go to 20 times higher energies. And we were assured that we could get the money through the government.

Kevles: There were a number of published statements along about 1945 and '46 that the reason we wanted those large accelerators was to investigate nuclear structure, not fundamental particles.

Wilson: I think there is an important thing to say here. One thing is that I think we had a false sense of what was known about nuclei. We had become familiar with neutrons; we had become familiar with resonance phenomena; and Viki [Weisskopf] put all this together at the time in the theory of resonance phenomena,[28] the widths and the separation of the levels. It all seemed to be just right there without anything more; with a few more experiments we felt very optimistic that we would roughly understand everything about

28. V. Weisskopf, H. Feshbach, D. C. Peaslee, "On the Scattering and Absorption of Particles by Atomic Nuclei," *Phys. Rev., 71,* 145 (August 1946).

nuclear phenomena. [In addition,] we probably did associate nuclear physics somewhat with wartime physics. There was a tendency to go on to the important problem in nuclear physics, which was to determine what the force law was.

Morrison: Bethe's lectures were given at G.E. in early '46 and very late '45—the winter of '45-'46—and they were sort of an introduction to nuclear theory and they were three-quarters or seven-eighths on the theory of the deuteron and all its manifestations. So this was the stuff that was missing almost entirely during the war and only a small bit put in for nuclear structure and saturation coming at the end. The real point was: Could this "hydrogen atom" be studied? It was a very complicated matter. And could meson theory be built up to account for nuclear forces? I think that was a major function.

Weisskopf: Maybe one should say that the terminology of elementary particle physics as distinct from nuclear physics occurred much later. Up to about, I would say, '52 or '55, the two fields were considered the same. For example, if one spoke about building multimillion volt or billion-volt machines for nuclear physics, what one meant was to get mesons. Nowadays one would say these were elementary particles.

Morrison: Aren't you really making the same distinction? There were two directions to go. One was toward the structural and phenomenological complexities of the nucleus, and the other was toward a more and more refined ultimate analysis of matter. They were both regarded as nuclear physics because the number of particles was not known to be any greater than one not-really-understood meson, plus neutron, proton, electron, and neutrino.

Telegdi: Excuse me. After the war, the idea was that if you built bigger and bigger accelerators, you would learn more and more about nuclear forces.

Wilson: You always started talking about nuclear forces, and the mesons came into this because they were the exchange particles.

Telegdi: One never heard the remark at that time that one would spend a great deal of time studying the mesons that would be produced.

Weiner: Then it did not represent any change. Because in the '30s the talk was of nuclear forces, and you are saying that this was a consistent thread. It is an old issue being tackled in a new way.

Weisskopf: On the other hand, let us not talk the other down too much. For example, there is a difference. When Fermi made that historic decision twice in his life—he first went from atomic to nuclear and then from nuclear to elementary particles, which is what we just heard—it meant something. It was only expressed in different words. There were definitely two directions to go after the war. One was to study the complexity of nuclear structure, and the other was to go down to the structure of nuclear forces, and that led into particle physics.

Merton: Is the historian of physics of this period then to conclude that you really had the same process reduplicating itself—the shift from atomic physics into nuclear, and from nuclear to elementary particles—and that the sense of identity of a special field follows long after it is under way? Even the terminology to designate it seems to follow. On the other hand, one has a sense of a definite shift in the publications and the documentary records. Now, the two accounts don't jibe. There were apparently some physicists making a shift as Fermi did.

Telegdi: I think it is a matter of vocabulary. I think that these words which designate fields arise only when the fields have already grown to a certain amount of independence. Right? We are doing some experiments now which will be called—I don't know—quark physics or some yet-unheard-of name five years from now and be classified as such, and you are not adjusting your terminology necessarily to a shift of your own tendencies.

Merton: Very good. That is the reason I raised the question. The emergence of new fields in other disciplines (if I dare stray beyond physics), even specialties in the social sciences, undergo this same process of the emergence of new interests that flow from work previously done and then there is a sense of recognition, a sense of identity. Now, one of the historical questions, which might be of interest, is this: At what point do you start to recruit the people into this new field? You have the pioneers who before they know it have moved into the new frontier. But at what point do the 90%, perhaps 95%, of the physicists, who have not yet been discussed here—not the frontiersmen—when do they start? When do the juniors start saying, "This is the new promised land."

Morrison: It comes automatically. They get Ph.D.'s.

Merton: Let us stay with that. That is one part of the reconstruction process. Now, what about these no-longer-youngsters, the old men of 35 in physics, who again are not the frontier. They are doing their solid pedestrian work, and I dare say that a bibliography of work in nuclear physics contains many of their papers. Do they make the shift? They can't make it as new Ph.D.'s. Do they represent, so to say, the remnants who stay in the preceding field? Is this the way it works?

Weisskopf: I think a majority stay in the preceding field because the preceding field is still growing and growing and growing.

Merton: I think someone this morning suggested the concept in this democracy of ours—the concept of aristocracy. And what you are really reconstructing is a process whereby not the self-defined aristocrats but the historically defined aristocrats...

Weisskopf: They are self-defined all right. [laughter]

Wilson: I think that is extreme because you can take the University of Wisconsin where there had been old accelerators before the physicists returned from war service. Those people would have regarded themselves as aristocrats because of the instruments they had and because of the decision not to build high-energy machines there. You saw them going off into very special work.

Livingston: May I come back to my central theme briefly? The theme is that the machines did come along and came along fast with McMillan's new ideas, and that attracted the experimentalists. They were originally nuclear physicists, low-energy physicists, who went right on into this new field and used the new machines, developed new fields, and one after another these machines were built. Berkeley got the first mesons with the 184-inch machine. At Brookhaven the cosmotron came in, and that opened another field of strange particles and associated production; and then the bevatron, and so on. But one by one, as the big tools became available, people came into the field, utilized it, and then the young ones began to come in with them.

Weisskopf: But it was a minority. Let us not forget it. Ninety per cent of the physicists stayed in nuclear structure, and nuclear structure expanded and expanded, not only scientifically but it also became part of astronomy. It became more complicated. It was of industrial importance for reactor construction.

Wilson: I would challenge those figures unless you actually made them. I do not believe that 90% actually stayed in. I think that some people, those, for example, who stayed at the large laboratories, stayed in low-energy classical nuclear physics.

Segrè: I think that the number of nuclear physicists now is much larger than the number of particle physicists.

Wilson: That is quite true.

Segrè: And in '46 you can imagine how much larger it was because the particle physicists were very few by definition.

Wilson: Nuclear physicists were still very few. I still think that there were roughly comparable numbers.

Morrison: One remark here. Just as the neutron in chain reaction has expanding devices, if you recall, the dilution by expansion is very important. There are many more physicists today than there were in 1945.

Segrè: There is one thing that we have not emphasized enough, namely, the part of cosmic rays in particle physics because particle physics existed before the machines existed, but it was called cosmic rays.

Merton: If I may try to reformulate what I think I have been hearing, it runs something like this: First, you are agreed that this is a question that can be investigated. There is some spread of opinion as to how the figures will fall out, but in principle it can be investigated. You can take a sample of physicists, arrange them in terms of their universities of origin, particularly in terms of age (I think this would be one of the most instructive questions) and begin to see who migrates from one field to another at which time. For example, you are suggesting the possibility that as you get these shifts in the prestige of fields within physics among physicists, you have a self-confirming process going on. That is to say, if what you are saying is true, it suggests something about those who remain. But this conception that you are evolving here is one that can be investigated, and I wanted to go back to the question Mr. Telegdi introduced this morning that seemed to me to be running through your entire account. It has not been investigated. And that is the age structure of the community of physics and the differences in behavior at various historical intervals in these three decades. You have implied a great deal about that.

Wilson: Let me say something about that. At the end of the war, the people who went into these laboratories were people roughly aged 35. There was a missing generation that had not been going to school who immediately as students populated all of the universities. They came out roughly five years later and became active. At that time you saw a very strong group.

Tuve: May I suggest just one word of caution? Because it was an applied science, there are a great many engineers, and you are going to have to decide in advance whether you are going to include them as nuclear physicists or not. But the pile design people had large numbers of people by the end of the war. If you talk only about the university community, that is one thing. But if you start talking about all the people associated with calculations about neutrons and nuclear physics, you have another group. Let me point out that we have, perhaps purposefully, for more than 20 years, confused science and technology for ulterior reasons, I think. But this does affect any kind of statistics that you try to make.

Merton: Mr. Tuve, I am indebted to you. My revised suggestion: no sample, total population; but the population would be generated by bibliography. That is to say, it is not those who hold positions in universities; it is those who have in some sense been doing something in the field—as witnessed by publications—and it would not be a total accounting of all those who may be teaching nuclear physics.

Tuve: You must remember that in nuclear physics most of the work for something like ten years was classified. It was later declassified.

Wilson: But I think the list that you start with is a prewar list. This is a finite list. You could trace those people through. Then you have to add names and eventually become statistical.

Weisskopf: There is one thing I feel definitely uncomfortable about. There is a value judgment here. The people who went into particle physics are people of a very definite turn of mind. They are people who like those fundamental simple problems and do not like complicated problems. I mean this in a very special way because problems in particle physics are also complicated. But some people do not like many-body problems, for example, they do not like problems of diversity. This is a kind of attitude. In other words, if one takes this point of view, then all the other physics simply goes down the drain—like solid-state physics and low energy nuclear physics. Today's nuclear structure has this same aspect that solid-state physics had, which means these are problems of complicated structures, so it is a different attitude of mind that leads people in that direction.

Merton: I want to make clear that in using this shorthand I was alluding to the question of finding out which men stayed in the field and which moved into the other field. We would then see what kinds of men they are—not in terms of their IQ—but where they are, at what ages, where they were trained and so on. The intent is to reconstruct the actual process, not to judge it.

Segrè: There is another thing to watch, and that is the tremendous influx in nuclear physics of chemists. There were many people who were chemists and became nuclear physicists, maybe still calling themselves chemists—I don't know.

Weiner: The Division of Nuclear Chemistry at the Lawrence Radiation Laboratory today is what nuclear physics was in the '30s.

Telegdi: I have two remarks. One I made when I read the report that was sent to us [study paper] and saw the eminent people who were being interviewed. It occurred to me that you should perhaps find a few cases of the opposite vintage. I mean people who left nuclear physics for one reason or another and ask them why they decided to get out of it. I think Felix Bloch at one time was interested in nuclear physics. At the end of the war he left nuclear physics on his own decision.

Segrè: Bloch was primarily in solid state, all his life.

Telegdi: No, I can show you to your great surprise a few papers by Bloch on nuclear structure. And I do not believe he measured the magnetic moment of the neutron for the purposes of solid state [physics]. If it is a bad example, it does not detract from the idea. To me it is a good example of a very eminent man who stayed away from particles and all those things.

The other thing is that I believe Professor Weisskopf is greatly embarrassed with respect to certain notions, but that does not necessarily detract from the truthful existence of such notions. There are several notions that I could mention that embarrass me, but many are possibly true and known to everybody. So therefore although he may be embarrassed about the Europeanization of American physics, nevertheless many people believe that such an influence exists. Although he may be embarrassed about the notion of aristocracy within the physics community, nevertheless I believe that notion exists. I believe what Professor Wilson said, that the path-finders, the leaders, go on into new areas and automatically get graduate students. The students then go into these fields, first, because they are graduate students with a limited amount of choice, and second, because sometimes the faculties have the most outstanding men. And we are back to the concept of aristocracy. But there is a feedback.

If some field of physics becomes very aristocratic and contains, for some reason or other the leading people, then the students ultimately will switch to those universities where these fields are cultivated. Especially these days, with government support, students can very often choose where they wish to go.

Merton: You may remember Garrett Birkhoff's injunction this morning: [that] before we theorize too elaborately, let us get some of the primitive facts. And the beauty of what has been reviewed in the last 15 minutes is that this can be inquired into in an orderly, methodical way. The primitive procedures for doing so exist. I would only bore you if I were to describe them. And they can answer many of the implied questions which we haven't even put on the table. For example, with regard to the oral interviews: Shouldn't you have interviews with men who have lived through the same period but who were not great contributors? How does all this look to them from the bottom of the hill? This is part of the history. The traditional history of science, centuries back, in the nature of the case had to be written largely about the elite, and that was difficult enough. Here we are talking about history in our own time; there is no need to perpetuate that crime against historical scholarship.

Telegdi: Yes. Exactly what I was proposing.

Holton: There is a very interesting mechanism at work with those graduate students who are the ones who are thrown directly onto the frontier. When a machine gets built on a campus, the facilities for handling graduate students almost double over-night. And that means that a very large number of students go into that field, but the recruiting mechanism for what happens to them afterwards hasn't changed. They cannot go on to new machines necessarily. Indeed, 60% to 70% may end up in industry. And I think our study should include what happens to these students who are thrown into this machine work and afterwards never do anything more in that field.

Merton: A parallel to this indicates that these patterns recur outside this one field we are examining. One of the great problems confronting the training of medical men in this country, who are drawn from abroad and who intend to return, is that they are trained with facilities that in the nature of the case will not be available at their home base. To that degree they are being mistrained, and we have been getting evidence of that over a period of time. So this, too, is part of the unwritten history being generated by the observations that have been made here.

Wheeler: In addition to these numbers that I think would be most interesting to get as you suggest, Professor Merton, there might be some words that could go with them: namely, what the applicants for National Research Council fellowships and other fellowships say at the moment they have gotten their Ph.D. degree or a few months before telling why they want to get a fellowship to do thus and such. I do not think there could be anything more revealing as to what motivates people, and they of all people are the most sensitive to the scientific currents of the time.

Merton: That is an adaptation of something which several of us are hoping to get under way. We have been given access to the archives of one of the great and distinguished foundations which has been granting fellowships not merely in the sciences but across the board of human knowledge for some 40-plus years. We will be sampling those since they run by now into scores of thousands of applications, to get not only their criteria (in their statement of plan, how they visualize the field they are in in each consecutive period of time) but also the criteria that the referees used in judging these men, which is the other side of the coin. In other words, what do the men who have made their mark in the field consider to be the hallmark of the comer in the various fields? We have no information as to how uniform these criteria are among the sciences or between the sciences and the humanities. So that if one could get access to these materials, Mr. Wheeler, there are two sources: the applications for admission to graduate schools, if those have been retained in some university centers, which could be systematically analyzed decade by decade to see how the field looked to the newcomer and secondly, the applications for postdoctorates.

Wilson: There is another point in a different direction. In speaking of postwar physics, again you have to talk about the war. A great effect in the large laboratories was the effect that being in those laboratories had on the men. I think that the physicists who came out of those laboratories were completely different men from the ones who had gone into them; they had a different idea of their relationship to the government. For example, they talked about setting up the AEC and had a great influence on how it was set up. This is something they had been deeply involved in. They were talking about setting up the National Science Foundation at that time. But the National Science Foundation and the AEC, in fact did not develop, and the ONR [Office of Naval Research] stepped in to disperse large funds. Of course that was very important and I suppose this factor would be easy to study.

I would say that having dealt with large and serious problems, with some success, the physicists were probably more arrogant. The idea of what they could do was probably considerably enhanced. And the fact that they did come out and have a different idea of the scope of the physics that would go on after the war than before—I think this is not an unimportant matter.

Holton: Isn't this the other side of the coin of the battle fatigue phenomenon you previously mentioned? Some people came out saying, "I want to go back to something which is 'pure', which is small scale. I've had enough of this. Take away your billion dollars."

Wilson: This is true, but I think that was momentarily battle fatigue.

Telegdi: On the other hand, a person like Szilard left nuclear physics by design, a complete decision. What I am saying is that there may be other cases which are less eminent.

Weiner: One of the reasons for his leaving—I think this was the case—was that he felt he could make more of a contribution to science in a field which did not involve a billion dollars, as Roberts put it, in machinery and that with ten dollars' worth of glassware, which he would probably break, he could do something more basic. Biology was his first love anyway.

Weisskopf: Szilard had another reason which I think ought to be mentioned here because it might be of some interest. Szilard made the definite statement that physics had gone out of the human range, so to speak. Physics, when it goes into particle physics, deals with phenomena that have very little to do with the human environment. It goes out of the human environment like astronomy dealing with faraway galaxies. Therefore, it is no longer so interesting and does not reach life.

Now, to my mind, it would be very interesting whether this kind of idea is widespread or not among the young people who choose a field of research.

Telegdi: Viki, there was another motivation. Szilard did not like technology, and I think the feeling that some of us who are in the mid-forties now have toward high-energy physics is similar to how he felt toward nuclear physics after the war. Even that amount of technology compared to what he used five years before was just too much for him.

Merton: There was a theme that ran through the discussion this morning and which found expression in some of the oral interviews reported in this document [study paper] that we might want to come back to for a moment. It is a theme which the physicists here find so self-evident that they can't bring themselves to examine it. And that theme was expressed over and over again in the phrase that practically every one of you has used: "It immediately became clear that—" and then you point to the next development. "There are basic problems to be solved here after the neutron," and so on. This turn of phrase indicating an absolute clarity, not of detail but of direction, as I say, turns up in the recorded interviews and turned up independently this morning here on many occasions, as today's transcript will show. This apparently represents a fundamental orientation of the physicist, but it is not clear to the outsider what this immediate clarity consists of.

Segrè: I will try to give you an answer. As soon as one had the Schrödinger equation, which one suspected or felt was exact as far as non-relativistic quantum mechanics was concerned, and felt that there was nothing else except Coulomb force in an atom, the atom was finished at least in principle. At the same time, it just happened that some new facts, experimental facts, were discovered in another field.

Merton: Are you sure that that answers the question?

Morrison: Let me give you another example, perhaps less general and abstract, because I think it is very much the same thing. When fission was discovered, within perhaps a week there was on the blackboard in Robert Oppenheimer's office a drawing—very bad, an execrable drawing—of a bomb. This was in January or February of 1939.

Merton: I rather suspected that that is what it was. What you are reporting is the living reality of science corresponding to what one was taught about the nature of science over a period of centuries as a conjoint enterprise, both empirical and rational. It is an empirical-rational complex, and if you have some understanding of it, you can see implications, whether they come from a new empirical finding or from a new conception.

Now, this raises the empirical question: How was this with the successively lower strata of physicists (not with you people for whom it was immediately clear)? How much of what is going on now that you find clearly pointing to the next steps is experienced as clear by those who are working in the vineyard but are not your students? Do they have the same degree of understanding? Now, if you worked with a small sample of physicists of various ages and so on, I think you could throw some light on what this aristocracy of scientific talent means. It is a composite of scientific tastes. It is a style of thought that leads immediately to implications (which I gather from the way you describe it often do not even have to be drawn because you have a kind of collective shorthand). When it is "clear," you don't finish the sentence. It is not necessary. All this raises many interesting questions, including the possibility that sometimes when sentences are not completed they collide in unexamined implications.

Telegdi: Excuse me. It seems to me that there are two classes of truly important experiments. Some are important because they are so obvious that they have never been done. They are obvious to everybody. I mean it does not take anybody special to say that to measure nuclear force it is best to put two nucleons against each other. That is very elementary. And the other great experiments like fission or something, which do not follow from any analytic source, are better described as the discoveries. I think that the aristocracy as such does not come only from recognizing the obvious and hence really important problems, because these are often synonymous. But many of the people who see these problems do not have the motivation, the strength, or the energy to go after them. I think the Columbia-Brookhaven neutrino experiment[29] is a patent example. Some low-grade theoretician in Bulgaria first proposed doing these things. First of all, even a high-grade theoretician in Bulgaria could not have done these experiments. Nor could they be done in the Soviet Union, where they were seriously considered by Pontecorvo.[30] In the United States these concepts were known to a number of people,[31] but there were only a small number of people with tremendous energy and dedication who decided to do the experiments. And I think, again, that this may be the sin of the aristocracy. It became obvious that neutrino experiments should be done. But the fact that something becomes obvious and will become important once it is done still means that there is a selecting out of the people who go after it.

Merton: May I just quote from an unpublished paper by my silent colleague across the table, Dr. Zuckerman? Some years ago, she engaged in an interview inquiry with practically all American Nobel laureates in science.

On the basis of those materials, she tried to find out the characteristics of both the crowned and uncrowned Nobel Laureates. Given such a numerically

29. G. Danby, J.M. Gaillard, K. Govlianos, L.M. Lederman, N. Mistry, M. Schwartz, J. Steinberger, "Observation of High Energy Neutrino Reactions and the Existence of Two Kinds of Neutrinos," *Phys. Rev. Lett., 9,* 1, 36 (July 1962). Experiment ran from October 1961 to June 1962 at Brookhaven on the AGS machine. M. Schwartz, "The Neutrinos," *Smithsonian Institution Annual Report* 1963, 359.

30. B. Pontecorvo, "Electron and Muon Neutrinos," *J. Exptl. Theoret. Phys.* (USSR) *37,* 1751 (December 1959).

31. M. Schwartz "Feasibility of Using High Energy Neutrinos to Study the Weak Interactions;" and T.D. Lee, C.N. Yang, "Theoretical Discussion on Possible High-Energy Neutrino Experiments," *Phys. Rev. Lett., 4,* 6, 306 (February 1960). Written without the knowledge of Pontecorvo's work, these references deal with the experimental feasibility of such a search. The concept of two neutrinos was proposed by many people, i.e., E. Konopinski, H. Mahmuud, J. Schwinger, I. Kawakami, M. Konuma, S.A. Bludman, S. Oneda, J.C. Pati, K. Nishijima. See *Phys. Rev. Lett., 9, op. cit.*

scarce value as the Prize, this should mean that one could assemble the same number of uncrowned Nobel Laureates whose collective contributions to the science of their day would be of the same order of magnitude as the collectivity of crowned Laureates. Everybody knows this but it is often forgotten. One of the major characteristics of this group, without giving them any psychological tests (which we could not have done even if we had wanted to) was described as tremendous ego strength. Now, ego strength was conceived in the following way: as a capacity for enduring continued frustration without manifest damage; a capacity to endure failure and to continue. It also involved what we called a degree of attractive arrogance, that is to say, that despite genuine humility, there was an overriding self-confidence, that if you had not hit on a significant problem, today or tomorrow or next year, sooner or later, you would. This quality is not characteristic of a cross-section of scientists, this capacity to know when you're doing pedestrian work while you're waiting for a serious problem to come along. Well, now, this whole complex does fit into the inquiry we are talking about.

It then becomes the question: How many of those who are not among the top level of physicists—in terms of achievement, not position—have these qualities of distinguishing between the pedestrian problems and the big problems, of being able to minimize their commitments to the pedestrian work (they do it but do not make it an absorbing concern) until they come upon a big problem. There may be a tie-in between the sociology and the psychology of the community of scientists.

Mendelsohn: Is there any distinction to be found here between the theoretical physicist and the experimentalist in this notion of "It became clear that..." or is there a difference in attitude? I mean, "It became clear that..." can easily be something which seems to come from a logical or mathematical statement, whereas it may not be an experimental statement.

Segrè: I would say the distinction is between the smart and dumb physicist, not between the theoretical and experimental.

Tuve: You have to grant also that there is historical continuity, so to speak, that when the time is ripe for a certain something to evolve, to be perceived, or to be discovered, it is. In general, most scientific discoveries or even ideas evolve independently in two or three places in the world. I think this should not be ignored. We are plotting, so to speak, the progress of the generic human mind. It is not simply dependent on the existence of a particular psychology in one person. Everyone paves the way for the next thing. Regarding this business of "It became clear that..." I think it is obvious whether you are experimentally inclined or theoretically concerned. Still, if you are interested in a certain subject and a sudden simplification occurs, it satisfies a very deep frustration you have previously suffered because you did not understand it. That is what we mean. My grocer would not have been at all concerned over discoveries that we were perhaps elated about. If he had been studying nuclear forces or something, he would have been elated.

Mendelsohn: If you were able to go back to some of these moments when "It became clear that...," has the phrase also been used for things which in a few years were found to be wrong?

Wilson: Oh, no. For example, at the time [early '50s] Fermi began to study phase shifts and a great deal of phase shift analysis went on. Everybody would say it had become the thing to do.

Telegdi: It did not take Fermi to do that.

Wilson: Maybe not, but, nevertheless, it was clear to all physicists that they should be studying phase shift. They did. All the theorists made analyses of these phase shifts. Nevertheless, it was a black period in nuclear physics.

Tuve: Things are not really wrong. Very often they are inadequately or insufficiently expressed, and they will have an entirely different real statement ten years later.

Wilson: I think by "not clear" you mean controversial. By "clear," you mean noncontroversial.

Merton: Over and beyond that, of course, there is the opposite situation, where matters are anything but clear. The false starts, the trial runs that don't pan out and so on. But at any given time there is a degree of consensus about the next direction.

Telegdi: I would like to make a few remarks. One is that it is true that "It immediately became clear that..." and hence you built a machine to make mesons and do the phase shifts. However, I think you will find on the record, in the post-war accelerator business, what people wanted the money for. I think the prospectus of the Chicago site was written by Fermi personally. You will find that the things that made those machines lively were totally unconnected with the proposal, and the technologies with which these machines were exploited did not exist. The Chicago cyclotron was built without scintillation counters. It is mysterious how the technology always grew to match the accelerator. The real payoffs invariably have not come from the "It is clear that..." type of issue.

The other thing I was going to say is just in a jocular vein. I have two friends who are theoretical physicists. One is a little bit more famous than the other, and one day discussing the less famous, he said, "You see, the difference between you and him is that he works only on things that he understands perfectly, and you work only on things that you don't understand at all." I mean some of us want to work on mysteries and other people want to clarify an already reasonably clear issue.

Wilson: There is one thing that probably has not been adequately covered, that is, the changing style in physics. That is an important consideration. We talk about the relationship between the experimenter and theoretician as though it were the same in 1930 as it was in 1935, as it was during the war, as it was immediately afterward, as it is today. And it is just constantly changing. That is a question that might be explored.

Merton: Is there any way that you can conceive of getting at that?

Wilson: I think that as the field matures, it is bound to happen. You could get at it, yes. For example, if you look again at the student's record, I suspect you would find that in this day and age there is a bias toward the theoretically trained student, chosen probably because of his mathematical ability;

whereas there was an easier separation, say, before the war, in terms of the tinkerer who wanted to become a physicist.

Wheeler: No Robert Wood could get into graduate school today.

Wilson: Yes, there was such a thing as a tinkerer who wanted to become a physicist, and he became an experimentalist. Nowadays I think there is much more of a selection of the mathematically inclined physicist. And you see that reflected in the way that experiments are being done today, which are much more remote from the tinkering. Somehow it is very difficult to see the physicist's connection except as an entrepreneur to the work.

Weisskopf: You consider this as detrimental?

Wilson: No, I make no comments except that I think it is a change. One could follow that as an historian, and it ought to be followed. You would find orders-of-magnitude difference in the kind of people who are doing physics, the way they are doing physics today, and the way they were doing physics 30 years ago. You would see a continuous change, a continuous development.

Holton: And also possibly a splitting up into different styles. You see, the "It is obvious that..." phenomenon has a certain implication, namely, it is that particular commonality of intuiting what is so obvious that pulls the group together. One remembers when Einstein explained why he did not go into mathematics although he had such good teachers and found it so fascinating. He said in his autobiographical notes that he discovered he could not distinguish between the important and the unimportant things in mathematics, whereas in physics he had the feeling that there are some things which are very important and some things which are not at all important. And that synchronizes a certain group of people who can talk to each other in shorthand. What they have in common is this matrix. Now, as long as it is a relatively small community, one can imagine just a few of these matrices in the community; but as it gets very much larger, different things will be obvious to different groups within the community as a whole. So style changes also might not be merely from one style to another but, let us say, from three styles to ten or more styles.

Weisskopf: I wish that that were true, but I am afraid it is not. That is probably one of the weaknesses of modern physics. If there were ten styles, I would be happy, but I think there is one style.

Wilson: One has gone from the highly individualistic style to something that you might call a managerial revolution in physics.

Wheeler: The style, of course, is a function not only of time but of place. I hope that somebody will dredge up a record of a talk that Kramers gave more than once about 1952 called "Physics As It Is Done in Different Countries." Nobody I can think of would be more appropriate to speak about that than Kramers, as nowhere can one see a bigger difference than between style in his country, the Netherlands, and in the States. There, of all countries, following the herd is not the way physics is done. People stick to different fields and develop them without feeling that their field is inferior to somebody else's field.

Zuckerman: Would this process of selection that you mentioned earlier, of taking young people with mathematical facility, facilitate communication between the theorists and experimentalists?

Wheeler: Yes.

Zuckerman: Would that in part account for what you describe as a tendency now for more validations being sought by the experimentalists from the theorists for their work?

Wilson: It is much more complicated. Technological changes have occurred. Earlier an experimentalist had to be devilishly clever in devising experiments in which he had to do all of the work by himself. He had only a certain amount of time that he could spend in the shops, or he had assistants who could help him. He had to spend that time very wisely. If he were going to make an important contribution, do something of significance himself, he had to think for a long time in the manner of Rutherford's students. When they moved, they were very clever and did something. Now I think that the technology has developed as well as the resources that a particular man can put behind the experimental work. Therefore you think, "Such-and-such has to be done. It's clear that such-and-such has to be done." Well, then, in a sense all you have to do, instead of asking for $100 as you did before, is ask for $6 million to do an almost trivial experiment. You can put all kinds of resources to use. In some sense it is just a measure of your ability as a manager of the technology of building, let's say, bubble chambers or building spark chambers. You can reach four or five solid angles. It is very hard to get much more than that [in an experiment]. You just have to put in a certain amount of intensity and you do that just by ordering the intensity and writing out an order for it. That is going to cost you money, but you do not mind, the money will be forthcoming. Maybe part of the experiment now is to get the money, but it is managerial.

Morrison: I would like to interject one thing, i.e., the relation of conceptual styles that took place is really interpersonal. It is very hard to isolate places in physics nowadays because of the excellent communications. After the war, in 1958 or '59, there was very little contact, minimal contact, at least on our side, between Russia and America. Yet I for one was rather surprised and somewhat disappointed on my first visit to a Russian laboratory because it looked so very much like what they were doing in Chicago. The only difference was that, while the sophistication on the accelerator side was, roughly speaking, equal, the detection side was very much less equal. No one has mentioned that at all. And I wonder what role fast electronics, cheap photo-multipliers and digital computation have to play in the postwar work. It is very important and has not been discussed at all, and it ought to be.

Session IV. Evaluation of Initial Work

Robert K. Merton, Chairman

Weiner:* The purpose of this initial round of interviews, which is not yet completed, was merely to get a first interpretation. We wanted to see if this conference was possible and to see if it would be possible to map the past development of an entire field of physics that is still developing. We made a deliberate attempt to get people's impressions, to find out how they thought—from their own partial perspective—things had developed. We knew that many of them would then be involved in a conference and in our future planning for historical research. So, there would be ample opportunity even to scrap the initial interviews and to start over on a better basis. It was for the purpose of developing good research techniques that these interviews were undertaken. Outlines weren't sent. We didn't have enough time, we didn't know enough, and we didn't have time to learn more. So we just went at it cold with very little preparation. Dr. Tuve was notified one day before that we were going to pop in on him, and it worked out adequately, I think.

Tuve: I carefully did not ask for a transcript. (Laughter) Mr. Chairman, I would like to ask one question which might help us a little here. In doing any problem I've always felt that almost a first essential is to have some notion of what you might consider an answer to be. I would like to ask here: What do you consider a proper outcome of this effort? I have a picture that what you're trying to do is to develop, by cooperation between the Academy, the sociologists, the historians, the Institute of Physics and the rest of us, a mechanism for acquiring, let us say, source materials and a depository of source materials for future generations or for more than one individual to study. So much of the discussion here seems to sound as though what we are trying to do is to achieve a single volume, *The History of Nuclear Physics in Our Time.* So, I would like to ask then: Which of these things are you trying to do? If we're expending all this effort, involving all these organizations, in order to achieve a volume on the history of nuclear physics in our time, then I think this is a bit of an exaggerated way to do it.

On the other hand, if you picture that you are trying to set up a pattern for the continuous deposition of materials for historical research concerning

*[An initial part of the evaluation session which is omitted here was concerned with the techniques of preparing for and conducting oral history interviews. After a brief discussion, Weiner was asked to comment.]

things, which may or may not be of ultimate interest, but far larger than any one study, then I think our approach to it, our willingness to have crazy things remain a part of the record, is just like our lives. We have all done many things which we would not do again, each of us. Nevertheless, they are a part of a human record. If you look at it in that sense, then a certain amount of winnowing is necessary. You cannot carry a whole generation fully because real time is real time. It would take a whole generation to read it. But with a certain amount of winnowing, you experts in history, sociology, and so on, can do this in some reasonable fashion. You can set up a pattern which the academies can follow henceforth and which physicists and perhaps later chemists and others might be expected to follow: Please make it a pattern that when you're involved in a lively subject, you talk into a tape-recorder and send it to a central depository or some such thing.

Notice, we have done this in the form of publication for generations. I take it you're trying to add another more modern dimension to the type of historical materials that previously have been available—perhaps to make up for the fact that we no longer write letters to each other. Could you clarify the goals?

Weiner: I have very little to add. I think that you've very well described the purposes of this project and our long-range goals.

Tuve: Then the aim is to establish a voluminous depository of many many different kinds of things, not to acquire a history of nuclear physics.

Weiner: It is the recognition that the history and the sociology of physics must be based on documentary source materials and that the interpretation of the historical actors can make some sense out of these materials and provide a cohesive tissue. That is one function of it. Another is that it can lead us to some materials that we may never have thought of. And another is that it may raise questions to interest historians and sociologists, to show that this would be a fruitful field for their inquiry. A further function is that involving physicists themselves in this activity alerts them to the very thing you mentioned, their historical responsibility. When they are doing something that they think may be of possible historical interest, they will preserve the records and make sure there's a permanent place for them.

Smith: By the time they know it's important, it's often too late.

Cohen: May I add something to that? While I agree with you absolutely and I think Dr. Tuve has given us a very brilliant statement of our aims, it seems to me that there's perhaps one contradiction in what you said; that is, that the complete complementarity that you imply is not genuine. For example, it is not really possible to collect source materials on a useful basis without at the same time having some feeling for the history of the subject. As you yourself have said, if you went to every individual who was connected with the development and got him to put it all on tape, no individual thereafter would ever have the time to listen to or read it. So the winnowing process and the process of selection involves a growing understanding. It seemed to us as we were thinking about this that at least for this first conference we would keep all of our materials in excessive detail, in a sense, so that we would be able to see in future projects where one might act more efficiently, what gaps we weren't quite aware of, or things we did that were unnecessary, and so forth. The collecting of materials, for the history that we ourselves are engaged in, is a new art. We're exploring it as we go along, and I take it that one of the chief reasons we wanted to have this conference, in addition to the collecting of materials, is in a sense to see how the different opinions and views that you've been expressing here are related to the views that come out of the materials we collect.

I think, for example, that something Professor Segrè said was extremely fascinating. We had played for us a record of Rutherford transformed onto a tape. And what Professor Segrè said was that it didn't sound to him the way it had seemed to him that Rutherford spoke; I believe he said Rutherford's voice was one octave lower. Now, this, it seems to me, is really very, very important. We all have the experience of saying things and hearing them come back to us when we've had to make recordings for radio or television and being somewhat shocked: "Did I say that? Is that how my voice sounds?" This is a terrible problem when we have the actors within the great drama of physics in which we live providing materials.

All I'm saying in a nutshell is: we're doing both. We're not trying to write the books, but we're trying to understand the subject and make sure that we're providing proper materials assembled in a meaningful order to be accessible to future generations.

Tuve: My hope would be that perhaps through this project, these joint projects with the Academy and so on, there might be a fairly clear assignment of where the initiative should lie for selected things that should be done. I do not believe that the men who are doing the work should be involved in the initiative function. That ought to be in a department like yours or in a group of the Institute of Physics.

Merton: Mr. Tuve, I would convert your formulation into a declarative statement because it was too good to be merely a question. It was a good, crisp formulation, I think, of what this group is trying to achieve.

Now, I come to an actual question. In order to get some sense of whether that works well, wouldn't it be useful to try to write a book or two on the basis of this material and see what the outcome is? If it turns out that all of these labors do not result in some kind of reconstruction that is superior to efforts of the past, it's all a promissory note. There's no checking on it. Now, that doesn't assume that there's a book that tries to cover the whole of the history of nuclear physics, but that simply thinking of the raw materials really doesn't subject this to enough of a test. It may have an internal coherence, but you don't know how it actually works out. Now, would you accept that as a possible suggestion?

Tuve: I am inclined to be a little doubtful. By what criterion do you decide that a book is good? It certainly wouldn't be concerned with sales records. Personally, I have the feeling that history must age a bit. So I would be quite willing to contribute effort to something which did not have a visible truth in our epoch.

Merton: I think this is the second time that there has been a unanimity of shaking heads and, unfortunately, in the same sidewise fashion. So I'll call on the headshakers. Mr. Wilson.

Wilson: There have been experiments in this area. There's a book by Hewlett and Anderson on the development of atomic energy. *The New World,*[32] I guess it was called.

Hewlett and Anderson actually did use some of these techniques. They tried to talk to all the participants. They did taped interviews, and then they eventually came out with a volume. They also took the point of view that a certain length of time should have elapsed before it became history. I think their criterion was 15 years, which is pretty short. On the other hand, I think that the book they did write was one that you find a lot of disagreement about but which most people probably regard as being somewhat better than no book.

Telegdi: I would say that this effort by Hewlett and Anderson is a bad example, because I think that there is more than one kind of history. I mean, the history of physics for the physicist is a topic for which we hardly need historians. In fact, that it was done by physicists is obvious.

But I mean that all during this discussion—this is perhaps what makes it interesting to us physicists—there are the non-physicists: the sociologists and professional historians. And that history [involving historians] would be an entirely different one if you want to write about physics at the time of Heinrich Hertz. You can't go and interview Heinrich Hertz. But even if you could, you cannot interview his underlings who were less successful. You can't get the flavor of their motivations. It seems to me that if you want to create this repository of material, which has psychological and sociological overtones, then you must change the style with respect to this kind of inquiry in the fashion that has been mentioned: interview people who have been very successful, people who have been less successful, and people who have left the field, that kind of repository, I think, is reasonably novel. It is novel to me as a nonhistorian. And that task you can accomplish because the entire spread of participants in this stage of physics is alive or a bit of the representative fraction is alive. Whether you make a book or not is immaterial. I entirely agree with Dr. Tuve concerning who is to judge: other historians, the physicists themselves? I don't know.

32. R. G. Hewlett, O. E. Anderson, Jr., *A History of the United States Atomic Energy Commission,* Vol. 1, *The New World 1939/1946;* Vol. 2, *Atomic Shield 1947/1952* (University Park: Pennsylvania State University Press, 1962 and 1969).

Kuhn: Let me say in reference to what is to be done with this data and is it for a book: I see no reason why somebody or some group—though on the whole I think this sort of thing is better done by individuals—shouldn't be working on a book at this stage of the game. But I think as soon as one starts talking about a book, one talks about the manner in which the interviews may fit into and give little bits and pieces for it, not of the job of doing the research for the book, which is not even fractionally done when one has interviewed virtually all of the people one is going to.

And speaking very personally, as a number of you know, I spent something over three years, most of it devoted to interviewing people who had participated in the development of quantum mechanics. For various reasons there were not as many of them as there would be in the case of nuclear physics: partly because there are more physicists, and partly because the developments happened earlier and there were not so many of them around. I found myself at the end of that time with a lot of fascinating information but not ready to write an article yet because I had then to turn around and start acting like a historian. I had to read lots of papers by the people who were alive but almost none by the people who were dead, and I had read them through a man's career rather horizontally, year by year and in a lot of other ways. I'm getting to that now; I'm beginning to do it. A book will come. A lot of books will come out of this.

But I think of oral interviewing—and I think this is the way Hewlett and Anderson used it also—as one of the ways of gathering data in the absence of letters and other sorts of things. I may add that where we've got them, letters are more useful than interviews, though there are certain sorts of information you get in interviews that never show up in letters because the people who wrote the letters took that for granted.

But my general feeling is that the interviews have got to be looked at in the way that Mr. Tuve suggests, that is, as long-term sources of data with tangential relations to a lot of future books and not in themselves as direct steps toward a book. That being the case, I express a slight malaise, which I perhaps have no right to, both because I am quite aware, as Charles Weiner has already said, that these [interviews] were to be exploratory with an eye to issues that might be raised. Also, I have not seen the interviews. Nevertheless, I get the feeling that a good deal of what has gone on so far has been more explicitly oriented to the questions one would ultimately like to answer than I think it probably ought to be in the long run. That is, the most useful information that comes through an interview in my experience is always the most personal. Now, by the most personal I do not mean internally, deeply sensitive, emotional issues, but I mean "I used to chat with So-and-So and I can remember what he said when . . . ." If the man is a theoretician, that is how I find out something about the relation between experimentalists and theoreticians. I think it is only very late in an interview, and not in the first one, that one gets around to asking the questions to which one's own research is going to lead one to a conclusion of what were the relations between theoreticians and experimentalists. I think you ask: "Who did you use to chat with? Who was at the table in the Berkeley Faculty Club when the physicists got together and what did you talk about?" You build up a much more day-to day biographical experience.

Wheeler: I was only going to add another comment in this direction in connection with the fact that we have discussed pure physics and applied physics; we've discussed pure history of science. Maybe it would be not out of place to say a few words about applied history of science. Of course I know nobody in physics (least of all the colleagues I see around the table) wants to orient his work in physics by what ultimate use it is going to be. And I certainly do not want to suggest that our colleagues in the history of science should orient their work by how it will be put to use. Nevertheless, I do feel that there are some uses that could be brought out that might have some bearing—direct or indirect—on what one does. And I think of three applications that I'll call 1, 10 and 100, number 1, number 2 and number 3.

The first one is the item of morale. What is it that makes the difference between a young man going to one laboratory where things really go wonderfully, and his developing into a really productive scientist, whereas the same young man if steered to another laboratory will never get anywhere? That factor of morale, whatever it is, if one could sort something like that out of this and write even the shortest article, even ten pages, I think it would be a terribly useful payoff from this work.

The second item is the 10, the 10 battles in nuclear physics that were most decisive in shaping the subject. If one could have on each of those just 10 pages on what the issues were, where the lively discussions took place, who said this and who said that, it would be useful. Perhaps the word "battle" is the wrong word because it implies a clashing of personalities. I do not mean that; I mean a clashing of points of view or ideas. But to sort out the 10 battles of nuclear physics I think would be very useful indeed to everybody who works in nuclear physics because so many young men coming into the field today have to fight these things out all over again among themselves. They don't realize they have been fought out in the past. They have not seen the cast of characters and seen this built up and that knocked down and finally this by unanimous agreement being the outcome. If they could see that, then they would not have to fight all these things over ingloriously and indecisively within themselves.

The third item, from 1 to 10 to 100, is the 100 peaks. Nothing disturbs me more than the fact that so many schools around this country have a beat-down look about them in the world of physics. You go in the front door and you know that that place is not doing much in physics, and yet you know that there are great areas of physics that are not being explored because nobody's taken, I think, the trouble to sort out the different areas of physics clearly enough and produce motivations enough so that University A will work in one thing, University B can find high-pressure work and can be the best in the country in high-pressure work. The idea that a young fellow can come along at a school and find, "By George, the people at my school are the best in the world in this field of physics; physics is an exciting field to go into; I am going to go into it in this school of mine even though it isn't Harvard or Berkeley," One would capitalize I think in this way on the creativity of young people around the country in a way that we are just missing out on every day, right now, all the time, due to the fact that we have not got these 100 peaks sorted out.

What is it that makes a difference between this country where there is so much following the herd in nuclear physics, let's say, and moving out of spectroscopy, versus the Netherlands, as I've mentioned earlier, where people are much more inclined to say, "The hell with this following the herd, I am going to work on this."? If we could sort out these facts. Well, we all know that in the Netherlands when people set the budget for the year the physicists get together—it's a small enough country—around a table like this and one says, "I would like to work on such-and-such." Another says, "I'd like to work on such-and-such." They collectively draw up the budget for physics in the Netherlands and they put it up to the government so that the subject is automatically divided up and, so to speak, the separation and the avoidance of following the herd is automatically taken care of.

Well, if one could sort out of this history of physics something that would help guide us to the 100 peaks and the 10 battles and the primary question of "morale," I think we would have three great payoffs of applied history of science.

Caws: We have been talking about history and sociology and not so much about the philosophy of science. It occurs to me that maybe the interviews as they have been structured are not being used to the best advantage with respect to the things that might conceivably be of philosophical interest. In particular, I think that the interviews might be used to generate data as well as to collect them, and I mean to generate in a rigorous sense—that is to say, to bring out things that even the person interviewed had not thought of.

Professor Segrè mentioned that some people keep diaries but not all; and that some people write letters but not all; and that some people introspect about their feelings about physical theory but not all. There have been some classical cases of people who did so in a rather charming fashion. Eddington somewhere has been quoted as saying that he thought of the electron as a little hard red ball and the proton as a slightly larger black ball. I don't think this is the level on which one ought to do it. All the same, it seems to me that when questions arise, for example, as they arose this morning in Professor Weisskopf's remarks about people's nervousness vis-à-vis phenomenological theories and so on, there might be questions that could be asked which would cause physicists to reflect on how they were relating to the theory at the time and cause them to have to say things which would be true but which, however, they had not formulated for themselves because they didn't happen to have that sort of habit of formulating such propositions themselves.

Weiner: I think in every interview, wherever there was an opportunity, we asked broad-brushed questions, which perhaps is the weakness of the first round of interviews. But I think there were reasons for doing it this way. And at the same time we asked for interpretations. The problem is, taking Tom Kuhn's criticism into consideration, that very often these well-defined, philosophical questions, are what you're after as an end result. And he is cautioning us to stay away from that sort of thing. We did throw many of them in in this case.

Caws: I didn't really mean end result questions but simply to stimulate the creation of certain data which were latent in the mind of somebody who was important.

Weiner: I think the things that were said today by Professors Morrison, Weisskopf and a number of others were in the interviews too. They contain some real reflections on what approach one takes to physics, how one approaches a problem, the style of work, and what makes one thing seem more important at one time than another. These are in the interviews so far.

Kuhn: I would just like to avoid any possible confusion. I myself don't see the conflict between what Peter Caws just said and what I have been saying earlier at all in the sense that I think Charles Weiner is hearing it. That is, again, I think one would not set up one's questions in order to start out "How did you think about it?" But we're trying to draw out the sort of thing which might not otherwise be forthcoming along with the course of individual recapitulation of events as experienced. It seems to me it is a very important sort of thing to do, so I would be all for trying to evoke this kind of information where it would come in. I would not set up the questionnaire in such a way that would make it look like its immediate object. Our experience generally was that to the extent that you get these things into the framework of individual biography, you get the richest sort of information and the sort also that you are least likely to be able to get from other sources. Individual biography does not have to mean going back to the beginning and a full account of family life, though sometimes if you get a man who will act as though he were doing his autobiography for you, when you get the next parts, he'll give you far more information than he would have otherwise. But a personal accounting of "things as I experienced them" with these things stuck in at key points by the interviewer seems to me the most likely source to give you the sort of data from which you're going to be able to examine other questions, including, I think, the very important questions that John Wheeler is raising. I would very much hope that this sort of enterprise could help considerably with them. I would only caution against the frontal approach, the frontal attack upon them directly in the interview situation.

Holton: I was much impressed by what Mr. Kuhn has said. I would like to say that even though in the long run the results may be unforeseen, we must have a great number of possible outcomes that we cannot describe now, if people are worried about publishing books. I do, however, want the idea kept open that pretty soon we can have some essays, some manpower studies, some chronology. In what day and what month did something get published, and when was it talked about? What did the graduate students do for their theses and what did they do later on in various representative lives and so forth? That's one [possibility].

Another is a series of essays which I think have been suggesting themselves to different participants. In my case, just making a random list of them, the essays that I would like to see a student working on, or perhaps work on myself, would be "The Experimentalist in the History of Physics in the U.S." or "Model Makers Compared with Fundamentalists," or "Where Do The Young Experimenters Go Afterwards?" or "Rival Models Working in Parallel for Several Years—How Do Their Proponents Communicate?" or "The Roles of the Emigrés and Their Incubation Period Abroad," or "Differences in Styles." I have a list of 12 people whose papers are simply not interchangeable; they have different styles of work and one can see it in what they're interested in, and so forth. Well, these are short essays which do not stand in any official way for the output, but which I think are the type of thing one had better get out fast in order to get the right use and high morale of any project like this to work.

Telegdi: Two questions: First, is there any cross-referencing of these interviews in the sense that Professor Segrè recalls that in June 1931 somebody told him such-and-such. Do you ever interview the guy who told him such-and-such so that they can refresh each other's memories? Many of these things occurred 30 years ago and, you know, everybody's memory is selective.

The second question is that it seems to me that the sense of gravity concerning nucleon particle physics is, at the present moment, on this side of the ocean. There are still many people alive in Europe who in lesser or larger roles participated in crucial developments. Perhaps they stopped contributing, but we're talking about anomalous alpha particle scattering and such things. Some of those people are alive. Rutherford's personal technician just died the year we had the Rutherford memorial. There are other people in Germany and so forth. Are these people going to be interviewed? And I think that since they're old, they should be gotten at sooner rather than later.

Weiner: This initial work was an experiment to see if it was worth doing these things and if it was worth continuing.

Telegdi: There's no doubt about that.

Weiner: And if the feeling is that we're on the right track and that, with the sharpening of techniques and defining of goals, we should continue, then I think we should mount a full offensive and be specific about what we really have to look into. The discussion today will provide the basis not only for the studies that Gerry Holton mentioned and for the interviews of people mentioned earlier (from the top, middle and lower ranks and from a wider geographical distribution) but for many other studies.

Weisskopf: The question is: What shall be done in the future? First of all, I second Charlie's remark. I think that the interviews so far have been useful and that what we did today has been useful to quite a great extent. The question is: What should one do in the future? I mean, apart from more interviews, interviews abroad (this is all more or less obvious) and also interviews of people who were not the most successful. I seem to come back to what I said once: the confrontation of a certain group of interviewees seems to be very important. Of course, this needs a lot of preparation, not only on your part but also on the part of the interviewees because they have to read the other interviews.

Merton: As the traffic officer of the afternoon, I declare there have been no bad collisions; there has been a concert of purpose and rapid movement and, quite seriously, a focus that had evolved out of the morning session became increasingly evident in the afternoon and must certainly make the chairman of the committee, Professor Cohen, very happy at today's developments and leave us looking forward to the next stage.

Could I now ask not for a vote of confidence or dismay but for two senses of the group. The first is on the question which is the concluding question of this conference: Is there tolerable agreement on the value of proceeding with this enterprise, with all of the ramifications that have been reviewed in the last six hours? Is there a sense of the group that can be reported back? Or is there a great diversity of opinion?

Wheeler: I must say, just speaking for myself selfishly, that there is no way I can get more help in trying to write a little survey of nuclear models than right here, just what I've gotten today. This picture of the history of nuclear physics is something that I have never had a chance to get all in one brief compass as we have here now. So from one man's point of view, I have a feeling that one can't do the job right without exactly this sort of thing going on.

Merton: Do we allow Mr. Wheeler to speak eloquently for all of us?

Telegdi: You can say that there's "intolerable agreement."

Segrè: I would like to say that this tape is going to be as interesting as, and probably more interesting than, many of the interviews.

All: Hear! Hear!

[ADJOURNMENT]

Enrico Fermi and Niels Bohr on the Appian Way, Rome, 1931. (Photo by Samuel A. Goudsmit, Segrè Collection, AIP Niels Bohr Library.)

Above: Left to right, Walter Heitler, Robert A. Millikan, R. H. Fowler, Marie Curie, Werner Heisenberg, Léon Brillouin at October 1931 Rome conference on nuclear physics. (Segrè Collection, AIP Niels Bohr Library.)

Below: Participants in the April 1932 conference at Niels Bohr's Institute of Theoretical Physics in Copenhagen. Seated in the first row are Léon Brillouin (left), Lise Meitner and Paul Ehrenfest. Seated behind and to the right of Ehrenfest is H. A. Kramers. The first six people, from the left, standing along the wall are Werner Heisenberg, Piet Hein, Niels Bohr, Léon Rosenfeld, Max Delbrück and Felix Bloch. Seated second from the right in the last row is P. A. M. Dirac, with R. H. Fowler on his right. Other visitors to Copenhagen in the group include Walter Heitler, Karl von Weizsäcker, Guido Beck and C. G. Darwin. (Photo: Niels Bohr Institute, Copenhagen.)

At the Cavendish Laboratory.
Above: Ernest Rutherford with J. A. Ratcliffe in 1935. (Photo by C. E. Wynn-Williams.)
Below: John Cockcroft (left) and George Gamow in 1933. (Photo, by K. T. Bainbridge, from AIP Niels Bohr Library.)

Above: M. Stanley Livingston and Ernest O. Lawrence in front of 27-inch cyclotron at Berkeley in 1934. (Photo: Lawrence Radiation Laboratory.)
Below: Robert Oppenheimer and Gregory Breit in 1946. (Photo: *Physics Today.)*

Proceedings of the

American Institute of Physics - American Academy of Arts and Sciences

Second Exploratory Conference on the History of Nuclear Physics

May 18, 19, 1969, Brookline, Mass.

THE ROLE OF THEORY
IN THE
DEVELOPMENT OF NUCLEAR PHYSICS

Participants at May 1969 Conference. (Photos by Heka.)

PARTICIPANTS — May 1969 Conference

Hans A. Bethe, Department of Physics, Cornell University

Joan Bromberg, Department of Social Sciences, Polytechnic Institute of Brooklyn

Barbara Buck, Department of History of Science, Harvard University

Peter A. Caws, Department of Philosophy, Hunter College of the CUNY

I. Bernard Cohen, Department of History of Science, Harvard University

Maurice Goldhaber, Brookhaven National Laboratory

Elspeth B. Hart, Center for History of Physics, American Institute of Physics

Erwin N. Hiebert, Department of History of Science, University of Wisconsin

Gerald Holton, Department of Physics, Harvard University

Martin J. Klein, Department of History of Science, Yale University

Robert K. Merton, Department of Sociology, Columbia University

Philip A. Morrison, Department of Physics, Massachusetts Institute of Technology

Emilio G. Segrè, Department of Physics, University of California, Berkeley

Robert Serber, Department of Physics, Columbia University

Dudley Shapere, Department of Philosophy, University of Chicago

Lewis Slack, American Institute of Physics

Henry G. Small, Center for History of Physics, American Institute of Physics

Valentine L. Telegdi, Department of Physics, University of Chicago

George E. Uhlenbeck, The Rockefeller University

John Voss, American Academy of Arts and Sciences

Charles Weiner, Center for History of Physics, American Institute of Physics

Victor F. Weisskopf, Department of Physics, Massachusetts Institute of Technology

John A. Wheeler, Department of Physics, Princeton University

Harriet Zuckerman, Bureau of Applied Social Research, Columbia University

Affiliations shown are those at the time of the 1969 Conference

Session I. Prewar Nuclear Theory

Charles Weiner, Chairman

Weiner: We thought that the first topic to discuss would be the general question of nuclear forces. It is of special interest because the search for the nuclear force seems to have been a consistent thread, which hasn't reached its end. It accounted, we felt, from the results of interviews, for the motivation of many people entering the field in the first place.

I will ask Professor Bethe to offer the first comments on this. Let me say that this is pretty much what our style will be—comments and discussion. I think that all of us—the physicists, the historians, the philosophers, the sociologists—should feel free to intervene, to politely interrupt, to challenge statements, to, in fact, learn something. We're all here to learn, and I think the only way we can do it is by having this kind of exchange. I guess it is not necessary to say that, since I can't think of any group that would be less inhibited than this one.

And now I'll ask Professor Bethe to begin.

Bethe: The force concepts until Yukawa's suggestion were, of course, all empirical, and I don't believe that we thought this would be the last word. On the other hand, I was certainly much more optimistic about the usefulness of the empirical force concept. And I don't agree with Emilio Segrè [who said in his interview] that "the study of the nucleus from first principles was already petering out by '34, '35."

If you consider first principles these various forces—Majorana, Heisenberg, etc.—were only beginning in '34, '35, and I was certainly convinced at the time that the things which prevented us from going farther with the exploitation of these concepts were not questions of principle but questions of mathematical difficulties—mathematical difficulties as applied on forces in a heavier nucleus.

And if I look back to my *Reviews of Modern Physics* articles, I believed very

Proceedings of the Conference on May 19, 1969. On the previous evening there was an informal dinner at the American Academy. Introductory remarks about the purposes of the Conference were followed by a talk by Hans Bethe in which he reminisced about the writing of his review articles on nuclear physics in 1936 and 1937.

strongly that if you could find the forces from scattering experiments and so on, and if you could then solve the mathematical problem of discussing the many-body nucleus in terms of these forces, that this would give the answer.

The first theory which started with first principles was Yukawa's theory and that only started in '35, but Segrè said [in his interview] "one hoped for very much out of Majorana exchange forces," and I presume he meant by that that this was a first principle theory.

Serber: At least one expected that the answer would be much more simple than it turned out to be. Yukawa was very convincing, and, of course, the reverse sign didn't seem reasonable as anything fundamental.

Uhlenbeck: Wasn't it also connected with the question of whether there was a repulsive force? There was for a while this question to get saturation you could do that by Majorana forces or you could introduce repulsive forces. And then, I still remember Fermi in his lectures at Ann Arbor, and he said, "Well, but of course, the nuclear force is not like molecular force. That couldn't be. That would be too simple. So we must try to do it without any repulsion and then by a combination of exchange forces and ordinary forces you could do that." One had somehow the feeling that there was a force and it should be simple, and it could not be like in a molecule, because then it would just be dull, so to speak.

Bromberg: Does this mean that between the period in '28 and '29 (when Gamow and Rutherford were talking about deducing nuclear forces ultimately from some sort of electromagnetic interaction) and the time of Yukawa people felt that they were working almost without any principles at all, that there was a kind of agnostic attitude?

Bethe: I felt there were difficulties and I think we will discuss that in a different context. I felt that difficulties had arisen from the idea of having electrons in the nucleus and they were really tremendous difficulties in principle. So I certainly never thought that this could lead anywhere.

The Gamow-Rutherford idea of deducing nuclear forces from electromagnetic interactions was a blind alley, and never led anywhere. The Yukawa paper of

1935[1] seemed to give a fundamental explanation but gave no serious quantitative results for about 20 years. The important thing for nuclear physics was just midway in time between these two, in 1932, Chadwick's discovery of the neutron and Heisenberg's paper.[2] This is when nuclear physics really began humming.

Now the question was: what were the forces between the protons and neutrons? There was first a suggestion by Heisenberg which wasn't quite right, then by Majorana which was much better, but the forces which were introduced were quite empirical.

Uhlenbeck: But still it should be something like the Coulomb force. In the atom there was a Coulomb force, now there was an unknown force, but it must be something as simple. Therefore it should be monotonic and that was what everybody thought, at least until 1935.

Bethe: I wrote down in my *Reviews of Modern Physics* articles that it couldn't be that complicated. I agree with Uhlenbeck that throughout the 1930's (not only until 1935) we pictured nuclear forces as simple and monotonic.

Weisskopf: Yukawa was, as you said, the first and maybe the last, attempt for a long, long time to put nuclear forces on basic principles, and it is a tremendous thing. The Yukawa question has not yet been further developed. I think we all have this belief that somehow the Yukawa step was the first step which can be followed and which will finally lead us to a satisfactory answer.

Weiner: Did you feel that at the time? The Yukawa paper was 1935. Can you date it from that? Or was it that once Anderson, Neddermeyer, Street, and Stevenson,[3] had found a particle which corresponded to Yukawa's prediction,

1. Hideki Yukawa, "On the Interaction of Elementary Particles I," *Proceedings of the Physico-Mathematical Society of Japan* (3) 17, 48 (1935).
2. W. Heisenberg, "Uber den Bau der Atomkerne, 1-3," *Zeit f. Phys. 77,* 1 (1932); *78,* 156, (1932); *80,* 587 (1933).
3. C. Anderson and S. Neddermeyer, "Note on the Nature of Cosmic Ray Particles," *Phys. Rev., 51,* 884 (May 1937); C. Street and E. Stevenson, "Penetrating Corpuscular Component of the Cosmic Radiation," *Phys. Rev., 51,* 1005 (June 1937).

Yukawa's theory came to be of such central interest? What about 1935, '36, that intervening period?

Weisskopf: I can only speak for myself, and my memory may be wrong, but I believe at that time, I felt, as I expressed it, that this was the wave of the future, that it was really the beginning of something great. Do the others agree with me, or not?

Serber: Yes, there was very little interaction either way, the theory didn't affect the experiments, according to Carl Anderson, and I think the experiments didn't prevent people from being enthusiastic about Yukawa's idea immediately when they learned of it. There was one precursor, and it was incorrect, but people did try to think that the weak interaction at small distances would become strong and might be connected with the nuclear forces. I wrote a paper which Hans [Bethe] turned down for the *Physical Review.*

Segrè: Fermi also tried it, saw it was not right and didn't publish it, didn't say anything about it.

Caws: Was it habitual at this time to make a distinction in everyday work between first principles and empirical considerations? Is this an *ex post facto* view of the matter or was it something you were always very conscious of? Was there a very clear line between empirical questions, working very close to experimental results, and questions of principle perhaps rather remote from them? Did these two things run into one another, or were they quite separate?

Bethe: I personally didn't take Yukawa seriously until the experimental discoveries of Anderson and Neddermeyer.

Morrison: Which didn't, of course, discover it.

Bethe: Which didn't discover it, but which discovered something which might have been it. But then, after that, as far as I can recall, I made this distinction very definitely.

Segrè: I would like to try and see why I said these things. As far as I can remember, and really I may be quite wrong. I may be constructing these things in my mind, but we are now in 1932. So Yukawa is in the future and there isn't any talk about Yukawa at all. Although Yukawa had read the beta decay theory of Fermi which was really his inspiration and he says so in his paper. But apart from this, the neutron was discovered. And a little before the official discovery of the neutron, Majorana had recognized that there was a neutron in the sense that when he saw the Curie-Joliot recoils of the proton, he immediately said, "Oh, they have found the neutral proton and they don't even recognize it. They are stupid as usual." This was Majorana's style.

This, I am quite sure, is the thing I remember. It is not invention. Majorana immediately after that started to make a model of the nucleus with neutrons and protons. And I am pretty sure that he had his exchange forces because of the mass defect of the deuteron and the alpha particle even before he went to Leipzig, except that he absolutely refused to publish this stuff. And he refused because with his calculations—he said silly things in this sense, "I'm going to start to now calculate the mass defect of all the nuclei." He started and he arrived to, say, oxygen, and then he couldn't do it any further. And then he said, "that's no good, the whole thing is wrong." Then he was induced to publish his paper only when he went to Leipzig with Heisenberg.

This is between 1932 and 1934, '33 and so on. By that time, at least he wrote the calculation of mass defects and binding and so on, and it was out of fashion. Then there was the discovery of artificial radioactivity which diverted the whole strength into that other field. This is what happened.

Bethe: Yes, but you had other things to do. It doesn't mean that it was out of fashion generally in other places. Majorana theory was not all out of fashion, and was eagerly pursued, in other places, including the U.S., England and Germany.

Segrè: It was considered very skeptically in Rome, I'm sure of that, at least by Majorana.

Bethe: But he considered everything very skeptically, including his own work.

Wheeler: I might make a few comments on another aspect of the story. To put nuclear physics into perspective it might be worth recalling that at this time there was a great deal of activity on the theory of positive and negative electrons, quantum electrodynamics, including the Bethe-Heitler formula, pair production, and so on. And one recalls, that just before this time—before one had confirmation of quantum electrodynamics through the discovery of the positive electron by Anderson—Dirac had proposed that the proton was to be understood as connected with the electron theory, being the negative energy state, the state predicted by a hole in the sea of negative energy states. And one also could recall the fact that there was so much going on at that time in terms of vacuum polarization. The work of Dirac and Heisenberg was very sophisticated and beautiful. I asked Heisenberg four years ago how did it come about that he suddenly gave that all up, and he said that he had thought that up until that moment of the discovery of the neutron that electron theory was truly fundamental, that the electron was more important than any other particle in the scheme of things, a point of view which I suspect Dirac still holds to even today. But Heisenberg, at any rate, gave it up with dramatic decisiveness at the moment when the neutron was discovered and entered into nuclear physics.

But already at that moment one had this concept of a pair being created out of a vacuum, out of nothingness. I would suspect it must have made some contribution to the idea of forming a neutrino and an electron in the nucleus in the sense of giving a background by which a production process could take place.

Bethe: This is really a different subject. I don't think that the people I talked to at the time, that is after '32, would have said that the neutrinos and beta decay and so on had any direct connection with the problem of nuclear forces. One might try this idea of stronger forces, that's true, but it never got anywhere.

Wheeler: Yes, this concept of exchange forces involving neutrinos somehow being the origin of nuclear forces didn't work, but one was desperately trying to stay in the stream of the history of molecular forces. You talked about how molecular forces differed from nuclear forces. Yet there was still this idea that molecular forces, though complex, had been successfully explained in simple terms, shapes of electron orbitals. There was this great hope that one could apply this same idea to nuclear forces. And, of course, one tried first one track and then another, tried electrons, tried neutrinos, and found that they simply wouldn't work.

But coming back—I'm just trying to put the subject of nuclear forces in a broader context—I think it would be very interesting to see what people's hopes were at this time, as distinguished from what was actually published, by looking at fellowship proposals, where a man tells, not what he has done, but what he hopes he can do. I would suspect that one would find that many young men expressed the hope that if one could only find the nuclear force, some new deep interaction, this would open the door to everything. And looking back on it from the perspective of today, even the perspective of the 1950's, nothing is more surprising than the discovery that the details of the forces don't matter so much, that nuclear structure can be worked out without knowing all the details of nuclear forces. The work is more in models and with the collective picture, than with details of the forces. With a few parameters that describe the important empirical features of the forces, one can reconstruct much of the observational data even though one is quite sure one doesn't know the deeper nature of the force.

Weiner: I'd like to comment briefly on the suggestion about fellowship proposals as a source of people's motivations. We are starting to look into the National Research Council Fellowships in the United States and we are just in the preliminary stages of it. It's possible to find out what people were interested in, but I rather suspect that these personal motivations don't get written into the proposals, especially when the motivations are more of a speculative nature, more of a hope that this is what you will find. And so I think the proposals would tend to be a little safer, even then, even in the thirties.

Serber: It is almost an empty question. That was the middle of the Depression. There were very few fellowships.

Goldhaber: But many applications.

Telegdi: I have a question. I know about all these attempts to make nuclear forces out of beta decay which failed. When Yukawa made his theory, had people already made such theories?

Weisskopf: Yes, Ivanenko, Tamm.

Telegdi: What I am saying here is what are the steps? There was a field theory of particles and somebody tried to make nuclear forces.

Uhlenbeck: It was $^1/r^5$. It is very hard to get it by higher derivatives to make it stronger and they even went to second order because there was an idea of a Russian who said that if you sum it all up then it becomes convergent and then it will become strong. I still remember strongly when Yukawa came and suddenly said there was another particle; then it seemed to me, that's too easy. With another particle, you can do anything of course. So until I saw the experimental discovery of the meson, I didn't think much of Yukawa.

Wheeler: It would be very interesting to look at the degree of skepticism as a function of time about the Yukawa forces.

Weisskopf: But you must admit this is a matter of memory. You people may have a better memory. I just cannot remember the time of Yukawa before the discovery of the meson.

Collective Chorus: Neither can I.

Weisskopf: As I said before, I was impressed by this first step into the real fundamental question of nuclear forces because of the identity of the range with the mass of the meson, but I just see to my horror that that was my interpretation. This is a very strange thing.

Weiner: Then you change your testimony from earlier, as far as the immediate impact of it. It has telescoped a little bit.

Collective Chorus: Right. Yes.

Serber: People weren't used to looking at the Japanese journals at that time. In fact, it was Oppenheimer and I who wrote a letter to the *Physical Review* to point out Yukawa's contribution, and we were especially impressed because we had been worrying about cosmic rays and we knew about the hard component.[4] It either meant that high energy electrons didn't interact in the way they were supposed to, or else Yukawa opened up another explanation which seemed to us very convincing at the time.

Goldhaber: Which year was this?

Serber: That was immediately after the paper came out.

Goldhaber: After Yukawa's, about '35 or '36.

Weiner: 1935. How was it that you happened to be looking at Japanese journals.

Weisskopf: That was typically Oppenheimer.

Serber: I don't know how it happened. It had probably been mailed and we happened to look at it.

Goldhaber: Wasn't he disappointed that it was written in English?

Holton: Mr. Serber, did anybody take seriously Millikan's idea of the hard component being the electron? You remember that his book, *Cosmic Rays,* was dedicated to the apotheosis of the electron.

Bethe: I know I didn't. In fact, I wrote a paper with Arthur Compton in '34 saying that this just wasn't possible and that it ought to be protons.[5]

Segrè: There was a big discussion in Copenhagen in '37 on the hard component with Euler bringing about the new mesons.

4. R. Oppenheimer and R. Serber, "Note on the Nature of Cosmic-Ray Particles, "*Phys. Rev., 51,* 1113 (June 1937).
5. H. Bethe and A.H. Compton, "Composition of Cosmic Rays," *Nature, 134: 3393,* 734 (November 10, 1934).

Morrison: Heitler's book, the first edition was '36 or late '35, is a prelude to saying that in $137mc^2$ something goes wrong.

Telegdi: I wouldn't say that. I would say the book says that people thought it would go wrong, but look, I show you the shower theory works.

Mossison: Not in the first edition. It's before the shower theory. There are two books. The first edition is completely different from the second edition.[6] The last chapter in the first edition is all about that. There must be something wrong. The multiple processes are probably what do it, and there is a sign of multiple processes. The shower theory was not yet known—it was Carlson and Oppenheimer—and it was thought that it all happened at once, and then at 137 the borders would give you something diverse. The second edition completely reversed.

Telegdi: You are quite right. The second edition completely reversed. It shows the great triumph of electrodynamics.

Cohen: May I ask a question which your remark gives rise to? In this period of the mid-thirties, how much information came from books, journals, reprints and conversations? Phil [Morrison] has mentioned a book where you distinguished between editions; you mentioned a journal article but it isn't clear as to whether it was looking at the journal or a reprint; there were meetings which were held at Michigan, Copenhagen, Leipzig and so on. What were the chief sources of information?

Uhlenbeck: I would say journals, and there were so few of them.

Serber: It is also true that people used to send reprints in those days.

Weisskopf: Not preprints, reprints.

Serber: I'm sure that is what we saw of the Japanese.

6. W. Heitler, *The Quantum Theory of Radiation* (Oxford University Press, 1st edition, 1936; 2nd edition, 1945).

Weiner: What about letters, though? For example, the compound nucleus, which we might get into later, was described to people who really needed to know about it, many weeks before it appeared in journals, through letters.

Bethe: I think this a singularity. Bohr never wrote anything, if he could help it, and, on the other hand, there were always people at Bohr's Institute who travelled a lot, and so this was spread by word of mouth and by letters long before it was published.

Weisskopf: Or by early editions of galley proofs of Bohr's papers.

Weiner: Maybe several years before the paper actually appeared.

Weisskopf: This is not a joke.

Goldhaber: This would be good historical research—to find all the galley proofs of Bohr's papers, which will differ considerably.

Morrison: There were, I think, two books of importance for students, or three, counting articles in *P R* as books that were perhaps used—Heitler and Dirac.[7] These were very widely distributed and used. Everybody who had ever looked at books had a copy of Dirac, and I think many people in addition had a copy of Heitler. These were the books, and everything else was much less important.

Bethe: What about the Pauli *Handbuch* article?[8]

Morrison: Also extremely important, but not quite as up to date. It seemed more a classic thing. Dirac still seemed up to date.

Weiner: An interesting point that Dr. Segrè made—I don't know if it was heard—was that the *Handbuch* article was much too expensive, that people didn't have it. What about the libraries of the institutions, though?

7. P.A.M. Dirac, *The Principles of Quantum Mechanics* (Oxford University Press, 1st edition, 1930; 2nd edition, 1935; 3rd edition, 1947).
8. W. Pauli, "Quantentheorie," *Handbuch der Physik,* Vol. 23, 1-278 (1926).

Collective Chorus: Yes, they had it.

Wheeler: Would it be fair to say that nobody made a contribution to physics who worked in isolation, who didn't do some discussing?

Weisskopf: Yukawa is a good counter example.

Wheeler: Did he not discuss with Tomonaga at that time?

Weisskopf: I do not believe so. I think Tomonaga is a lot younger, but I am not sure about that.

Goldhaber: Tomonaga is still older.

Wheeler: And also with Nishina.

Morrison: He was also in Leipzig. By the way, the paper was certainly published in German and not in English.

Weiner: I think Henry Small has a question on forces.

Small: If we can get back to the force question, I would like to clarify this pre-Yukawa stage, which has been described as an empirical stage. Of all the competing theories—the Heisenberg approach, the Bartlett force, the Wigner, and Majorana—was there a general agreement as to which of these was the best, or was there considerable disagreement?

Morrison: I think that really is a wrong point of view. I'm speaking as a student at the time, but I suspect the teachers will agree. This was not regarded by anybody as being different theories. These were just different mathematical terms in a formula which is one and the same theory. Each ingenious algebraist found one way to make it more complicated and more general and to finally write the whole thing, expanding it in the complete set. It's like saying sine-x and sine 2x and sine 3x. It is much more like that than like three or four theories.

It is the development of one single idea with great mathematical generality and elegance, perhaps.

Bethe: I agree.

Segrè: Do you know if the story is true that I was told of how the Bartlett forces came about. There was a seminar and they said no exchange for ordinary force. Exchange of [position] coordinates only was Heisenberg. Exchange of coordinates and spin is Majorana. And then somebody asked, what is only exchange of spin, and Bartlett said, well, it's so, and this became the Bartlett force.

Serber: Bartlett, as a matter of fact, said it as a joke.

Weiner: Were you there at that time?

Serber: No, I wasn't, Bartlett told me that he had said it as a joke.

Holton: I would like to follow up what John Wheeler said which seems to me quite intriguing: that perhaps nobody or hardly anybody contributed who did not have a chance to discuss. If this is right, is there something in the dialectic itself of the problem which required it? It is wellknown that Bohr had to talk it out with someone in order to understand it fully. Is it an accident, or is it so complex? What is the reason for it?

Segrè: People are competitive and they want to show that they are smarter . . .

Bethe: I don't agree with the thesis. I think Yukawa, even if he had some other Japanese to talk to, certainly was far from the mainstream of the theory, and I think is quite different. Dirac never discussed anything with anybody and made a contribution.

Segrè: Neither did Majorana for that matter.

Weisskopf: Schwinger does not discuss.

Morrison: Yes, that's right. It is quite individual, isn't it?

Wheeler: Isn't there a communication of concern about what the problem is and about what is important that is "in the air?"

Bethe: Yukawa undoubtedly read the journals.

Morrison: He had just come from Leipzig. He certainly knew the gossip as of a year before.

Uhlenbeck: And there were only two journals, or three.

Cohen: What were they? I think I know, but I'd love to hear for the record.

Uhlenbeck: The *Zeitschrift,* of course, and the *Proceedings of the Royal Society,* and then, in the thirties, also the *Physical Review.*

Klein: Wasn't *Nature* still the medium for rapid publication in the thirties?

Goldhaber: Yes, especially for experimental work.

Bromberg: How did the *Comptes Rendus* stand?

Uhlenbeck: Nobody read it.

Segrè: People began to read it when the Joliots became active. Then they published every week. People would look to see if there is anything new from Joliot.

Weiner: Wasn't it typical to take a journal then, say *Physical Review* or *Comptes Rendus* or others, and as just pointed out, to look for the name of the person that you would think was worth listening to?

Uhlenbeck: In those days journals came out once a month. Not *Comptes Rendus,* but the *Zeitschrift* and the *Physical Review,* and you read them cover to cover. You read everything, you didn't just pick, as one does now.

Weisskopf: Don't forget. This is not exclusive. There were very important articles in the *Proceedings of the Physical Society,* in the *Physikalische Zeitschrift.*

Goldhaber: The *Cambridge Philosophical Society.* There were lots of marginal associations' journals.

Telegdi: I have always heard the rumor from Zurich that Stuckelberg, who after all invented so many things, also had the idea of the meson, but he was talked out of it. Has anyone else heard of this idea?

Weisskopf: It sounds plausible.

Telegdi: Supposedly he had the idea of a massive particle to explain short-range forces. It could well be. Some of you may have talked to him at the time.

Holton: Who would have talked him out of it?

Telegdi: Pauli.

Weiner: I would like to get back to a point that Phil Morrison made before, in answer to Henry Small's question, that in fact these were variations on a generally agreed upon theme. Now I gather you were referring to the period of the early thirties. I would like to know if this tendency of the variations on the central theme persisted throughout the thirties in terms of the approaches to forces.

Morrison: I think there is an over-reification of these names, Wigner, Bartlett, Heisenberg—that is all I meant to speak about—it has nothing to do really with the general theory of forces and how they arrived. So I really can't quite untangle what you're saying, Charles. It is just wrong to refer to these as four theories or two theories. It is all one conceptual doctrine with various mathematical formulations, and there might be many other integrals or sums with people's names connected with it, which you would never think of associating with theory. Let me say emphatically that theoretical physics has got this in it very much. It's like experimenters who have experimental designs and discoveries and also have instrumentation in the same way that theorists have conceptual frames that perhaps grow, or perhaps some great man, or some lucky hit, produces. And then they have a lot of instrumentation—people who make this into calculable materials, and very often we give names to those because they are ingenious devices, useful to refer to, and so on. And I think that is a better parallel than the parallel that they are very strange in theory—it is all seen as one. It had really nothing to do with theory as you are conceptually trying to analyze it. It was a technical matter.

Weiner: Then it would apply even to present concerns. It is a helpful guide to us in looking at it.

Serber: After Yukawa, people began to investigate different kinds of meson theory to get the different kinds of forces.

Bethe: But that's again one kind of theory with different names.

Morrison: There, at least, it wasn't a purely technical difference, but it was sort of a conceptual framework given, and then various physical possibilities outlined. There were a whole series of those.

Bethe: Also, the phenomenological forces could still have any shape, any dependence on distance. The only thing we knew was that they fell off very rapidly with distance, but it could be exponential or gaussian or what not. We could have given different names to those instead of to the different exchange characters.

Weisskopf: This shows that it is not really a theory. It is just a semi-empirical device.

Morrison: Exactly.

Uhlenbeck: Could one perhaps say that in the early days one thought that the nuclear forces were, so to say, the universal force of the same calibre as the Coulomb force in an atom. Then one said, couldn't one guess that force from the given facts. Now, already it was said in those times, that it was clearly impossible. Suppose that you didn't know the Coulomb force, and you would only know all the levels of iron. Could you have deduced the Coulomb force from that? That approach was clearly given up very soon, I think. And then it was clear that one must have an *a priori* theory for the force. Then, as I said, there was this beginning to deduce it from the beta decay theory, which didn't work out at all.

Bethe: I don't quite agree. I think there would be the desire to have an *a priori* theory anyway, but once we had scattering experiments, I think there was at least the possibility to deduce the phenomenological theory without an *a priori* theory from the scattering experiments.

Uhlenbeck: That was, of course, much better than the binding energies.

Bethe: That's right. A two-body problem in more than one level.

Caws: Would it be correct to say (and this is perhaps what is reflected in Professor Segrè's recollections) that, between the old principles that were available—Coulomb forces and the proton-electron nucleus and that kind of thing—and then the Yukawa, there was a period when there were no principles available from which anything intelligible could be deduced?

Morrison: All that were known had been tried and failed. Right.

Holton: What was the morale at that point? Did one feel one was on the threshold of something very new and great and startling, or does one feel that all the tools that one has available have come to an end and it is a disorganized period?

Wheeler: Breit was working very hard on interpreting nuclear reactions at this time and had a great deal of success because he focussed on things that were explainable, for example, the dependence of cross-sections on energy insofar as it depends on barrier penetration. There were other things, of course, that one couldn't explain—the selection of energy levels of nuclei. On the other hand, one should recall the work of Heisenberg and other people who had treated nuclei as built out of particles in terms of which there were empirical forces, and one could discuss binding of these nuclei. One didn't have a great deal of success, but still one felt one was on the right track.

As Hans Bethe remarked earlier, Wigner showed along these lines that one needed this force to have short range to account for the much greater binding for three particles than for two particles. So it really wasn't at all a hopeless enterprise.

Weisskopf: I think it depends on your attitude. One could distinguish between the fundamentalists and the practitioners, and, of course, a single individual may have both components.

I would answer your question to some extent positively. There was a certain discouragement in the thirties that nuclear physics cannot be dealt with in the same wonderful way in which the twenties had dealt with quantum mechanics in the spirit of Dirac's radiation theory, of Fermi's radiation theory, which was such a wonderful thing at that time. It was only a few years old but one could write fundamental equations and quantize Maxwell equations, and get out everything apart from a few infinities.

So I do believe—again, I can only speak for myself—that there was this feeling of frustration in the fundamental direction, but there was so much else to do that many of the fundamentalists became practitioners, if you want to use these words. Let me say it perhaps this way: there are two ways of approaching nuclear physics. One is if you want to do it by fundamental principles. The other is almost the opposite, as Johnny [Wheeler] was saying before, namely, trying to understand the experiments with as little as possible fundamental input. And that was the idea, I think, which the compound nucleus concept started. The compound nucleus concept of strong interaction, the nuclear temperature idea, and then the whole theory of nuclear reactions, the Bohr-Mottelson approach, these were all attempts to understand as much as possible with as little as possible fundamental assumption. This is the complementary approach to what physicists had done before, what physicists like to do starting with the fundamental. If you call this frustration, it was frustration.

But on the other hand, it was also exhilarating that you could explain so many things with so little input.

Segrè: There was so much going on that there was no time to be frustrated. There were slow neutrons, there were isomers, there was K-capture. I could list twenty discoveries that came. Every two months there was a new one.

Serber: But the door was still open. It wasn't frustration in the sense that one didn't see how one could answer the question, say, talking just of nuclear forces. As Hans [Bethe] said, the scattering was expected to give a very direct answer, and what was needed was, first of all, the scattering experiment, but then what was needed was much higher energies so one could get in and actually see the shape. That didn't come until later because the technical facilities didn't even exist. One couldn't be discouraged as long as one saw a way of answering the question.

Bethe: There are really three levels. I think Viki [Weisskopf] has described very well two levels, and Bob Serber has described an intermediate level, that is, you ask for a fundamental theory, such as meson theory, from our present point of view, and such as electrodynamics in the case of, let us say, atomic structure. And then you ask for an intermediate theory which starts on phenomenological forces and describes binding energies in terms of phenomenological forces. And then there is the purely phenomenological descriptive theory, such as compound nucleus, and Bohr-Mottelson, and so on. And these are three different levels.

There was progress certainly on the two latter levels all the time, phenomenological forces, and actual phenomena.

Segrè: What about the progress in finding new things that nobody expected?

Bethe: Finding new things, yes, and I think there was hope on all three levels.

Telegdi: I must say that perhaps there is another aspect. Is it not true that people were mathematically minded? Once Yukawa's theory came out they made infinitely many variations around that—chemists and those people—all these mesons with various spins, various charge theories, and so forth. When I became a student right after the war, the situation as seen—this is just a complementary view—was that, first of all, there were infinitely many meson theories that you could, as a student, study if you wished without any clear preference for any of them. And then, on the other hand, it wouldn't do you much good to study them because none of them gave a calculable result. This is roughly the picture that one got.

Bethe: At that time there was frustration, actually, until Powell's discovery, because nothing worked. There was so much choice. There were lots of contradictions.

Wheeler: When one speaks of the point of view of the 1930's, as Viki did, it is interesting to see how much things changed in the course of the 1930's. Just as one poor man's example, in 1934, at the spring meeting of the American Physical Society, I can remember my first year as a National Research Council Fellow.[9] I was 22, giving this paper at the Physical Society, and beating the drum about how you could get so much out of studying cross-section as a function of angle and energy, how much you could learn about the forces. And this was the great white hope of nuclear physics—to get the forces from studying the scattering. And in 1936, only two years later, here was the compound nucleus model of Bohr, and it was the direct opposite as Viki had stressed. You shouldn't hope to get anything about the forces this way.

Bethe: It depends on which system you study. If you study proton-proton scattering, you would still hope to get out the forces. If you study the scattering of neutrons by uranium, you wouldn't.

Wheeler: Exactly.

Caws: At the last meeting of this group I think it was Professor Weisskopf who told the story about somebody—Schrödinger perhaps, I'm not sure—who watched people going in and out of the swimming pool?

9. J. A. Wheeler, "Interaction between Alpha-Particles," *Phys. Rev., 45,* 746 (May 1934).

Weisskopf: It was Heisenberg.

Caws: Heisenberg. What I'm wondering is the applicability of this to the question of what is inside the nucleus and what is outside it?

Weisskopf: This is a different chapter.

Caws: I know it is an entirely different chapter. But was there a point at which—having only a small number of previously known particles at one's disposal and the expectation of finding the force relations between them—there was a kind of inhibition against new views of the nucleus? Was there something that had to be, as it were, discarded before progress was possible, a conceptual attitude toward the nucleus?

Weisskopf: Maybe I don't quite understand your question. I would say, no, but I am not sure whether I understand your question.

Let me first go back to this frustration angle. I am quite sure that this frustration was felt, for example, by Heisenberg. If you notice, Heisenberg did not participate in the work on the level which Hans has described, the purely empirical one where you would like to get something with a few nonfundamental assumptions. There were a number of physicists who just did not take part in this activity.

Uhlenbeck: Pauli also. He thought it was just like solid state. "Schmutzige Physik."

Weisskopf: Pauli is even a better example. And certainly they had that feeling of frustration that we are speaking of.

Holton: What made it filthy physics? What makes it schmutzige Physik?

Weisskopf: That you have to make assumptions which you cannot prove from first principles, that's all.

Goldhaber: It is not only that, I think. It is so complex and complicated that you cannot easily test your assumptions. Even if you make the correct assumptions, you cannot test them by bombarding lead with neutrons.

Holton: I think there must be something a bit more complex in addition. Can one narrow down this at all—the esthetic balance within Pauli, or within anyone that you know?

Weisskopf: I will try, but probably wrongly, to base it on that contrast of which I spoke. Pauli was not interested in explaining such phenomena as, for example, the energy distribution of the neutron in a reaction . . . .

Bethe: He was not a practitioner.

Weisskopf: . . . . and referring to the nuclear temperature concept. Because the nuclear temperature concept, for Pauli, did not give him a deeper insight into what is really interesting. Because Pauli—I hope he is not mad at me when I say that—Pauli is not interested in just one phenomenon as such. He is interested only in what this phenomenon teaches us about the fundamental laws of nature. And it is true that that kind of thing which we practitioners at that time did was to sort of explain experiments by simple concepts but not going into the deeper question of what it is really that makes the nucleus tick.

Bethe: Was Pauli interested in Yukawa?

Weisskopf: Yes, definitely, especially because it happened at the time where Pauli and I wrote the wave equation for this new particle.

Bethe: Were you aware, at that time, of Yukawa?

Weisskopf: That's a very good question. We were talking about a possible particle that may fit this, but I cannot tell now whether . . .

Morrison: I feel that a very useful substructure of this kind of record approach would be given us if we simply had a list of topics and dates of the whole sequence of commentarial papers Pauli writes about Wigner-Jordan, Pauli-Jordan, each one of which did something to make a field theory on either a new spin, pairs, or something like this. There were a whole series of them. And I think they formed, as far as I know, the principal theoretical guide for the next ten years.

Weiner: Was the Pauli paper 1934? This one: "On the Quantization of the Scalar Relativistic Wave Equation" [with V. Weisskopf, *Helvet. Phys. Acta.* 7:709, 1934] ?

Weisskopf: Yes, that's it. That was before Yukawa.

Morrison: But now Yukawa undoubtedly read that.

Weisskopf: Yes, but we did not connect it at all with the nucleus. But the idea of this paper was to have a different model of a theory of positive and negative charge, of charge symmetry, as one would say today. A different model of charge symmetry which doesn't have to use the concept of filling up the vacuum, the hole theory. We did talk, and this I remember very well, we did talk whether maybe there are particles like this . . .

Segrè: The alpha particles.

Weisskopf: At that time we knew something about the structure of the alpha particle. The alpha particle did not fit into our theory because if you go to interaction energies which are equal to the rest energy of the alpha particle, the alpha particle does no longer exist as an alpha particle. It does not really fit our theory.

Caws: This is really another form of my question: at what point were people first prepared to entertain the possibility of a particle outside the then canon of particles, or combinations thereof, like alpha particles?

Serber: It started with the positron. Once one got over that shock, you could expect anything.

Segrè: I'm quite sure, for instance, that people like Rossi, had suspicions of particles long long ago.

Bromberg: How long?

Segrè: I don't know, but I know, for instance, of experiments that were made to measure curvature and specific ionization long long ago by very primitive methods because of the showers, the triple coincidences in showers, and so on. One should really ask Rossi because he would really tell you something.

Weisskopf: He is not in town. He is in Rome.

Bethe: Well, there was Serber's remark about the shower theory. That was when? 1936?

Morrison: 1937 was Carlson-Oppenheimer.[10]

Segrè: The discovery of the showers must be around 1930, the triple coincidences. Now, soon after that, and I can't remember precisely when, I know that the cosmic ray people—Rossi, Bernardini, and so on—suspected that in this thing there was something which was neither electrons nor protons.

Wheeler: They always spoke of the penetrating component.

Segrè: And they didn't know, of course; they tried to find out and they couldn't.

Wheeler: Early in 1935, in Copenhagen, I can remember E. J. Williams discussing the evidence on these penetrating particles. There was always the question, of course, that many people had worried about: was electrodynamics correct in predicting electrons could not penetrate? Or could some of these penetrating particles be electrons? And then this discussion whether the Bethe-Heitler formula is correct. And then Bohr's concern to get a simple way to show by the most elementary arguments that the Bethe-Heitler treatment must be correct; and this giving one therefore much increased conviction that these particles simply could not be electrons. And this furnishing the background for the work of Anderson, and finally proving that these had to be mesons.

10. J.F. Carlson and R. Oppenheimer, "On Multiplicative Showers," *Phys. Rev., 51,* 220 (February 1937).

Caws: Among physicists in general, though, this business of accepting new particles intermediate in mass, and so on, must have spread rather slowly, since people were still trying fundamental explanations in terms of the already known particles. Is that correct? This is what I mean by inhibition—a line that has to be crossed, after which one entertains a whole spectrum of things where before there were few conceptual slots into which possible explanations could fit.

Bethe: That was certainly true of me. I resisted believing even the Anderson-Neddermeyer discovery, and I would have liked them to be much rather protons and anti-protons.

Weisskopf: Is it not so that before Yukawa the whole idea of a particle with intermediate mass did not exist? Am I right?

Segrè: No, you are wrong. The cosmic ray people believed in mesons really long before Yukawa.

Morrison: I don't think they believed. I would say it was much discussed at Berkeley.

Segrè: Or independently.

Bethe: What was wrong with protons for the cosmic ray people?

Morrison: What was wrong with the state of electrodynamics? I still think that was the most plausible thing until after Williams and so on. About '36 one began to be sure that it was probably right, and the shower theory then was a quantitative demonstration that it was right. And then we said there must be something odd and now we should try very hard to find e/m for these particles, and that, I think, was what gave Anderson and Neddermeyer the push.

Segrè: When were the experiments in which you measured drops and curvature in a cloud chamber?

Bethe: Nothing before Anderson and Neddermeyer. '37.

Holton: Did Street and Stevenson believe that this was an intermediate particle?

Bethe: They are a little later. In my opinion, it took their experiments, in addition to Anderson and Neddermeyer, [to show] that there really was an intermediate particle. Anderson and Neddermeyer had proved they were not electrons.

Segrè: Look, this is a convincing proof, but before this convincing proof, I wish I could ask some of the cosmic ray people.

Bethe: I remember that Rossi said way back, in 1930 or so, that these were not electrons, but that did not prove they were not protons. And that the primary cosmic rays could not be electrons, I think that he had been saying as long as I could remember him. Everything fitted protons and that was the big fight between Rossi and Compton on one side, and Millikan on the other side.

Weiner: That was resolved by '34 or so, wasn't it?

Segrè: By the East-West experiment.

Holton: But the Millikan book is '38, isn't it?[11]

Bethe: It was resolved, but Millikan never believed it.

Weiner: It was resolved. The Millikan book is another story.

Small: Can we talk a little bit about the difficulty that people were experiencing in electrodynamics at this time? Did this deter people from doing this sort of work, or was it an encouragement? You were speaking of the difficulty in the early thirties, and this comes out in many of the interviews that we have collected and in the letters, so I wonder if we could talk about that.

Weisskopf: There was a time when I did work on electrodynamics. I don't quite understand your question?

11. R.A. Millikan, *Cosmic Rays: Three Lectures, being the revision of the 1936 Page-Barbour Lectures of the University of Virginia and the 1937 John Joly Lectures of Trinity College Dublin.* (New York: The Macmillan Company, 1939).

Bethe: I don't quite understand the connection.

Weiner: Part of the thing that we have noticed is that theorists have tended to go in and out of fields in pursuit of certain problems or classes of problems, generally where there is a chance of success. And you find people flitting in and out of nuclear physics and electrodynamics during this period. You find an interaction of these people crossing back and forth. We want to explore the effect this flitting about had on the development of nuclear physics itself as a field. In other words, was there any steady core of theorists who stayed in nuclear physics? And, if so, was it because they were deterred from doing anything further in electrodynamics?

Weisskopf: It is again this fundamentalist versus practitioner angle. It is a matter of temperament. I can give you a list of people, starting with Heisenberg and Pauli, who never went into nuclear physics. And I can give you a list of people who sort of went back and forth like myself. And some people who didn't touch electrodynamics because of its problematics. And I personally think there were too many of those who stayed away, because if there were a little more people interested in it, I think we would have had the Schwinger-Feynman development in '34. We came pretty close to it. But there were just not enough of us, and we were not intelligent enough.

Wheeler: I might make one comment in this connection—the connection between nuclear physics and electrodynamics—the beauty of pair theory. The statements that we will be coming to later that electrons couldn't have anything to do inside nuclei because one couldn't localize them that way, always encountered my skepticism. Was there a way, I kept asking, in which nuclear forces could be explained in terms of positive and negative electrons because nuclear forces differ from electric forces, not by 20 powers of 10 but by 2 powers of 10, or 1 power of 10? So one could ask, could the radiative component of force be more important than the static component of the force?

With all the beautiful things in pair theory and all the differences from the ordinary static theory of interactions between particles, it seemed reasonable to explore whether the nucleus could be viewed as built up somehow of positive and negative electrons interacting. So this was always at the back of my mind and led to concerns about half advanced, half retarded interactions.

Thus, if radiative forces are to be dominant in the whole discussion, and overpowering the static forces, then one had to explain why radiation did not come out in large amounts. And so one had to go back to half advanced, half retarded, potentials. This led to all this question of relativistic electron theory, and then we finally were able to develop this further.

But the motivation in the first place of getting into that subject was to try to understand nuclear properties. I wouldn't hold to that now. I think that most of us now don't regard the electron as that fundamental.

Weisskopf: I don't understand the time scale here. You and Feynman worried about this when?

Wheeler: 1939.

Weisskopf: Yes, and after all, the problem of the electrons in the nucleus . . .

Wheeler: Most people regarded the problem of electrons in the nucleus as thoroughly disposed of much earlier.

Weisskopf: That is what I meant. But you didn't?

Wheeler: But I refused to think of this as reasonable. In fact, I always thought that the Yukawa theory of nuclear forces was much overrated. It seemed to me to be explaining something in terms of something that was itself a mystery.

Bethe: But I think this was an isolated opinion.

Wheeler: Yes, this was very much an isolated opinion.

Goldhaber: But very useful, as you point out.

Bethe: For something else, but not for this.

Goldhaber: It is a lesson for the historians.

Weisskopf: For that matter, he is right. The meson is an unexplained phenomenon. It may consist of two quarks.

Morrison: That is true of every phenomenon in physics, I must say.

Caws: This is what seems to me philosophically interesting. You have a tradition, at this time, of something more than a hundred years of a more or less worked out atomism. You have a tradition of maybe twenty or thirty years of understanding that there are sub-atomic particles, but these particles still have a kind of classical being—they are little lumps of stuff. And at some point or other, this rather simple-minded practical notion of particles disappears in favor of something much more abstract and complicated. And I wonder to what extent an inherited philosophical view—first of all as to the nature of the particle, and second as to the fundamental simplicity of the furniture of the world—prevented, if it did, the emergence of nuclear theories based on a quite different understanding of what the physical situation was.

Morrison: I don't think that is the case. I think the theories were absolutely continuous from the time of Fermi-Dirac radiation theory, with various brilliant complications to make it handle pairs and so on, but it was the same theory. This can be distinguished by two things which should be teased apart. The idea that a particle and a field were complicated complementary sort of things was sort of very clearly held. I think it was the ruling idea that this was the real understanding you need. The reluctance was now to choose many independent parameters—that is what you were saying—to describe the world. Having got a field theory, why do we need n-field theories.

I remember Bohr saying in 1937 and it impressing me very much as a student—I could barely understand what he was saying in more than one way—that we would not understand the small world in detail, that means all the particles, nuclear structure, and everything else, until we understood the large world. And there was no sign of that happening.

Caws: By the large world, he meant the universe?

Morrison: The cosmos. He hoped . . . He said that unless there is a context in which these can be connected, we won't understand anything. And, at one point, Bohr even came back to that later and said, "in all the fields together. There's no point in trying to study electrodynamics," and so on.

Bromberg: May I ask something really connected with that? What was the reaction in 1931, for example, when Bohr was saying that the electron only exists with respect to its charge in the nucleus but doesn't exist as a mechanical entity? Did that seem very strange at that time? This is in a speech he wrote up after the Rome 1931 Conference, which apparently he wrote up from the remarks he was making to people at the time.

Weisskopf: You must have taken a sentence out of context there.

Wheeler: I think there is a Faraday Lecture of Bohr at about that time in which he speaks of the fact that the number of particles is not a well-defined object. If you try to localize an electron too precisely you create pairs. So the charge is there, yes, but the number of particles, no.

Segrè: The trouble was even and odd.

Weiner: I think Henry Small has the quotation referred to.

Small: Bohr was speaking of beta decay in this context, and he said: "In this situation we are led to consider the capture or the expulsion of an electron by a nucleus plainly as an extinction or a creation, respectively, of the electron as a mechanical entity."[12] And this was dated 1931, at the Rome Conference.

Weisskopf: When was the paper by Fermi?

Segrè: December of '33.

Weisskopf: Bohr had the idea in the form of a qualitative formulation two years before Fermi.

12. N. Bohr, "Atomic Stability and Conservation Laws," *Convegno di Fisica Nucleare, Ottobre 1931,* 128 (Rome: Reale Accademia d'Italia, 1932).

Morrison: I don't think it is, because it would have had to do with nonconservation and angular momentum. It's the opposite. I think it is the same Bohr line that led to the idea of the statistical nature of conservation which was dead wrong.

Many Voices: It was much later.

Morrison: It was much later but I think it was the same idea . . . .

Goldhaber: That is a little too strong. It has one correct component but then when it was realized you need another particle . . .

Segrè: In fact, at that Rome Conference, there was really disagreement between the people who wanted a new particle and the neutrino to save everything, and Bohr.

Weisskopf: Was not Pauli involved in this discussion?

Goldhaber: But he was not at the meeting.

Weiner: Yes. According to Sam Goudsmit's recollections Pauli walked into the Rome meeting as Goudsmit was telling about Pauli's neutrino idea. Goudsmit's paper in the proceedings of the Rome meeting says, "At a meeting at Pasadena in June 1931, Pauli expressed the idea that there might exist a third type of elementary particle besides protons and electrons, namely 'neutrons'."[13]

Weisskopf: This reminds me of the picture we saw yesterday.

Weiner: That is another proof he was at the meeting.

Dr. Serber has to leave, within a half hour or so, to make a 2:00 faculty meeting of the Physics Department at Columbia, and I would like to ask a question now that sort of follows and that also relates to some of his experiences.

13. S. Goudsmit, "Present Difficulties in the Theory of Hyperfine Structure," *Convegno di Fisica Nucleare, Ottobre 1931,* 41 (Rome: Reale Accademia d'Italia, 1932).

Could you comment on the proton-proton scattering experiments and the validity of charge independence in the late thirties? This seems to have led into the continuation of postwar scattering experiments with renewed hope for getting something out of them.

Serber: Well, on the proton-proton scattering, the first experiments that I know of were Milt White's at Berkeley. They just established there was a proton-proton force within a factor, say, two or three of the proton-neutron force, that is, of no enormous differences in magnitude. It wasn't until the later experiments of Tuve gave sufficient accuracy that the force, at least in the singlet S state, turned out to be very close to the force between the neutron and proton. There was also the question of charge independence of symmetry. The evidence of the mirror nuclei again indicated an interchange of neutrons and protons, and that was done first by Lauritsen, Fowler and Delsasso in Pasadena.[14]

Weisskopf: When was that?

Serber: '35, maybe.

Well, my recollection is that it was Wigner, wasn't it, who proposed the isotopic spin as a conservation law?

Goldhaber: It was Heisenberg.

Bethe: No, Heisenberg introduced it as a good notation for transformation of neutrons into protons, and vice versa. It was Breit, Condon and Present; and Cassen and Condon,[15] wasn't it, who proposed the conservation law in the charge independence, and then it was Wigner who made so much of it in the systematics of nuclei in general.

14. W. A. Fowler, L. A. Delsasso and C. C. Lauritsen, "Radioactive Elements of Low Atomic Number," *Phys. Rev., 49,* 561 (April 1936).
15. G. Breit, E. U. Condon and R. D. Present, "Theory of Scattering Protons by Protons," *Phys. Rev., 50,* 825 (1936); B. Cassen and E. U. Condon, "On Nuclear Forces," *Phys. Rev., 50,* 846 (November 1936).

Weiner: This was '36?

Bethe: '36 was Breit, Condon, Present and Cassen, and '37 was Wigner.

Segrè: When was Wigner's big paper of all the isotopes and classification of all the multiplets?[16]

Morrison: 1937.

Telegdi: I have a remark which is very much related to what Professor Bethe said about Wigner's work. You know the same work that Wigner did was also done by Hund.[17] He worked with charge independence, also with spin independence, a very general abstract thing, and one day I asked Racah why he was working on the spectrum of iron, atomic spectra of this kind. He said, well, when I was a young man I read the papers by Wigner and Hund on the symmetry principles as applied to nuclear levels, and I studied that stuff, and I learned all the group theory that went with it, and then I wanted to apply it, but I discovered that there were no nuclear levels known to speak of, so I tried the same concepts back in atomic physics.

Serber: Wigner also, I believe, suggested that in nuclear reactions isotopic spin conservation would show up and affect the rate very much. And I believe that Oppenheimer and I found the first example of that in one of the light element reactions where different resonance levels had alpha widths of very different magnitudes. Then there were the experiments on nuclear forces before the War, and, of course, after the effective range theory was established.

Bethe: Breit had the main information on that already before the War.

Serber: I think Landau had done it before the War.

16. E. Wigner, "On the Consequences of the Symmetry of the Nuclear Hamiltonian on the Spectroscopy of Nuclei," *Phys. Rev., 51,* 106 (January 1937).
17. F. Hund, "Symmetrieeigenschaften der Kräfte in Atomkernen und Folgen für deren Zustände, insbesonders die Kerne bis zu 16 Teilchen," *Z. Phys., 105,* 202 (1937).

Bethe: Landau and Smorodinsky,[18] I don't know what the time was of that relative to Breit, but certainly Breit did all the hard work on showing that different shapes of potential would explain the same proton-proton scattering experiments equally well. He did not say, I think, explicitly that there were only two parameters, but when you looked at his curves, it was perfectly evident that there were only two parameters that the experiment gave you.

Telegdi: But it must have come to him as a disappointment because he had been encouraged by getting better and better measurements not realizing that he would get only two parameters out of it.

Weisskopf: What is the story about Landau and Smorodinsky? You developed the concept of effective range.

Bethe: Really it was Blatt who developed it.[19] I put it into a simpler form, but Blatt had the concept.

Weisskopf: Is that so?

Bethe: That is so, yes. My paper [20] is more understandable, but Blatt had it before me.

Weisskopf: And what was the role of Landau and Smorodinsky?

Bethe: They had done something close to it in proton-proton scattering. They didn't call it this, but they pointed out that there were only two parameters, and I think this was based on the Breit calculations. It was later than Breit's calculations, but earlier than Blatt and myself. And Breit hates them.

Serber: At that time, the knowledge of the forces in scattering really was very meager on anything except the S-state.

18. L Landau and J. Smorodinsky, "On the Theory of Scattering of Protons by Protons," *J. Phys., USSR, 8,* 154 (1944).
19. J. M. Blatt and J. D. Jackson, "On the Interpretation of Neutron-Proton Scattering Data by the Schwinger Variational Method," *Phys. Rev., 76,* 18 (July 1949).
20. H. Bethe, "Theory of the Effective Range in Nuclear Scattering," *Phys. Rev., 76,* 38 (July 1949).

Bethe: And only two parameters on the S-state.

Serber: And it wasn't until the Berkeley cyclotron gave neutrons of 90 million and protons in the 100 million volt range that any more really detailed information on the shape came out. I would say the principal thing that did come out from the initial analysis of that was that the long-range part of the force had the Yukawa range. The details though turned out to be very difficult. In fact, I remember we couldn't even establish the isotopic spin rule in any way uniquely. One could put it in and make it fit but one couldn't deduce it from the results.

Bethe: Are you referring to pre-War?

Serber: No, I'm referring to the analysis of the Berkeley results.

Segrè: What you mean really is that in Berkeley there were two steps, One, was when Robert Wilson, using the modified 37-inch cyclotron, which I think went up to 14 MeV, started to see the P-wave in proton-proton scattering. At that time, everybody said, if you see the P-wave, then you will know everything. Well, of course, that wasn't true.

And then when the 184-inch started to work, there were first the runs on neutron-proton—the neutrons were made by stripping—and I remember Lawrence, who had heard about nuclear exchange forces, going around with an ionization chamber, putting the chamber in front of his belly, and in back, and seeing the difference in ionization, saying, here is proof of the charge exchange forces—I remember this quite well—in the beam of neutrons.

But later, of course, there was the measurement of the angular distribution in proton-proton and at first great misgivings because everything was flat. Nobody knew why; how is it that we are certainly at energies sufficient to create waves beyond the S-waves and everything in the proton-proton angular distribution stays flat.

Bethe: You had the neutron-proton first, didn't you? And the proton-proton later?

Segrè: That's right.

Serber: Let me go on with what we've learned. And, as I say, there was a great difficulty that the cross-section came out too large and we resisted the idea of a repulsive core until Jastrow really forced us to . . . well, that was the only way we could get out of our troubles.

Weisskopf: That was much later. That was the middle-fifties.

Segrè: I wouldn't say middle fifties, probably very early-fifties, because by middle-fifties, we were already through with all the polarization and everything, the triple scattering, and everything under the sun.

Bethe: '49 or '50.

Morrison: '51 or '52.

Weisskopf: It was too late for Blatt and Weisskopf, I remember.[21]

Weiner: You should have lingered with your galleys.

Goldhaber: Like Bohr.

Weiner: I would like to ask, Dr. Serber, what you meant in your interview when you said: "It was clearly understood by that time [1946] that there were limits as to what you could find out with the energies of protons and neutrons that were available previously." You mentioned some of this, but what I would like to know is, if it was clearly understood by 1946, was it not clearly understood before the War? In other words, the intervening period when much of this physics was suspended didn't bring any new information to make it more clearly understood. Or did it?

Serber: We had the effective range theory . . .

21. J. M. Blatt and V. W. Weisskopf, *Theoretical Nuclear Physics* (New York: John Wiley & Sons, 1952).

Bethe: In 1949, maybe 1948.

Serber: The facts were, in fact, known that the scattering was insensitive to shape.

Bethe: The Landau-Breit calculations.

Serber: And also that you had practically no information on the odd and even momentum waves from the scattering and it was clear you had to go to higher energy to get any idea of the shape.

As the range was small, you need short enough wave lengths to analyze for the short-range structure.

Weiner: Did this happen in the immediate postwar period?

Bethe: I think it was realized by 1940. On the other hand, it was only in the postwar period that there was any chance to get ahead and to make a big cyclotron. It was deferred for lack of equipment.

Segrè: Wasn't the specific idea that you wanted to see whether the P-states, the first one that would come, were attractive or repulsive, and from this infer which kind of exchange it was? I think this was lurking.

Bethe: That was certainly one thing, but I believe that probably we were more ambitious. We knew that we ought to get to much higher energies like 100 Mev to get anything about the shape that was worthwhile. The wave length equal to the length of the force.

Segrè: But people were very concerned, at least let's see the P-wave and see whether it is attractive or repulsive.

Bethe: I don't dispute that, but that was not the only thing that was wanted.

Telegdi: There seemed to be another problem that was very popular when I started studying physics, which is the opposite of what you [are saying about the need] to have high energy to put neutrons and protons together. That was one way of getting it. And the other way which seemed to be very popular, at least the experimental idea, was to study the photo-disintegration of the deuteron. There were these papers by Schwinger and all these things which centered on the interest that if you could just get 10 MeV gammas or what have you, it would be a great thing because it would be twice as good as 10 MeV neutrons and protons. These switch patterns were done, these nuclear patterns, nuclear reactions. It was a very hot subject at that time, another way of getting nuclear forces.

Bethe: Yes, but I think one should emphasize there was a need for an accelerator to go much higher than 20 MeV, and this wasn't possible until the McMillan-Veksler invention.

Morrison: But the 187-inch.

Bethe: That would never have worked without FM. It was realized everywhere outside of Berkeley.

Holton: If I may ask a question while Dr. Serber is still here: it may be an optical illusion, but I have a feeling that when you look at the ages of the people in the field, that the curve shows a much narrower band-width than in other fields, like, let us say, the quantum mechanics development, or the relativity before that. In those pictures that we saw yesterday, the old Lorentz was on the wall, and the people in front were more or less of an age. Now, with the great exception of Bohr perhaps, there are none of these great statesmen in the background . . .

Segrè: Rutherford.

Holton: Well, Rutherford only up to a point, because he turned out to be very wrong, but as a stimulus for the young people . . .

Weisskopf: What time are you talking about?

Holton: I'm speaking about the early thirties.

Goldhaber: In which sense did Rutherford turn out to be very wrong?

Holton: In his guess that he thought the nucleus would turn out to be a very simple kind of thing. I'm not saying that we are having here someone to dismiss, but rather: is it an optical illusion that the group that was very vigorous was more or less of one age and that they were not tightly coupled in their imagination to a previous generation?

Weisskopf: I believe that this has something to do with the previous generation being what I call fundamentalist in theory. Rutherford did not have, at that time—I'm speaking of the mid-thirties—a big influence on the development of nuclear physics.

Bethe: That is correct, but he had had a great influence in stimulating the Cockcroft-Walton development, and he then did a lot of very important experimental work on deuteron reaction.

Weisskopf: This is correct, but maybe there is a point in it. Let me try to explain this thought. I think it is striking that the kind of physics that was successful in the second half of the thirties and then later on in nuclear physics was a very close connection with experiment—much closer than it was before. And it may be that kind of attitude had to be created by those younger peopel. Am I wrong? I don't know.

Bethe: I agree with that very much.

Goldhaber: Bohr had this attitude. Einstein did not have it so much.

Holton: This is one aspect; the other is that in other, earlier kinds of physics there is quite frequently evidence of strong national differences and national schools. Certainly, relativity was done differently in Germany, let's say, than in France, to a large degree. I wonder whether here again I'm correct in seeing homogenization, that the national differences appear to have gone largely away when one comes to nuclear physics, with perhaps the exception of the French isolation, to a degree, until the effect of Joliot-Curie experiments.

Weiner: I think you're wrong in the sense of the experimental aspects, anyway. There were certainly important national differences in the attitude toward large-scale industrial type engineering projects of the type that worked in Berkeley. When you talk with people, as we have done, in Italy about their experiences in building a cyclotron there; talk with Kowarski about how they got started in Paris; talk with Chadwick about the difficulties of getting started in England—it's a completely different culture, a different kind of matrix in terms of thinking and in terms of types of projects tackled . . .

Holton: That's what I meant when I referred to relativity. There I was thinking about the theoretical aspect, where you can correlate differences in national styles and in the theory that is being done.

Weiner: I was referring only to experiment, but theory was so closely linked with experiment.

Bethe: Even among the experiments, the Cockcroft-Walton was, after all, the very first that worked and gave results. It was a smaller enterprise than the Berkeley enterprises but it was an engineering enterprise.

Weiner: Well, the point is that they didn't escalate it.

Bethe: They didn't escalate it but they got the first artificially produced particles making a disintegration in the Cavendish. But they didn't escalate it, and this is important.

Weiner: By the way, an interesting point that has come up in our work is the intense criticism that Rutherford was apparently subjected to in the twenties for pursuing nuclear physics when it wasn't a field, when it wasn't defined as a field. He was criticized for taking groups of students and training them in nuclear physics when there were no clear ways for them to go. There was no vocational pipeline.

Klein: What was the source of the resistance?

Weiner: Colleagues in other fields in physics, people in England in particular, who felt that Rutherford was really derelict in his responsibility for going off on this tangent.

There is a statement in print[22] which I came across in preparation for this, and then I had the opportunity to explore this some more with Chadwick. And this is what I meant at the last meeting about a field beginning to take shape. Certainly the field existed for Rutherford, but it didn't for many of his colleagues.

Weisskopf: This is very interesting because it was also the time when Rutherford said that to ever get energy out of atoms is so much moonshine.

Weiner: That is really later. He said that in '33 or so, '32 or '33, whereas I am now talking about the period just after the creation of the new quantum theory.

Morrison: Of course, he had not always said that.

Weisskopf: About the energy—when did he say differently?

Morrison: Quite differently, during and before the First War, '12 to '15, in many talks. He talked about this conservation of energy, endless energy. After all, Rutherford was in nuclear physics since 1897 or 1898 steadily.

Segrè: What would a physicist who had started physics in England do to become a professor? There was no other opening, was there?

Cohen: The National Physics Laboratory took a few people, and the Navy took a few, but very few.

Goldhaber: Remember when you look at those photographs at the Cavendish in the twenties, the total number of students was one row or two rows, of six or seven each, so the total production of what you might call nuclear physicists

22. Marcus L. Oliphant in *Hommage à Lord Rutherford. Ceremonies du X Anniversaire de la Mort de Lord Rutherford,* 18 (Paris, 1947).

was very few—two or three a year. I think when I got my degree there were about two or three simultaneously getting degrees.

Morrison: One of the principal reasons for homogenity is the exponential rise in the number of people. Except that the people who are in the field are always very new in the field and therefore they tend to be young. Now, when the group is eight or ten, you see an older gentlemen, who is, after all, the boss, a singularity. Then you say, well, they're spreading all over. But as soon as there are many young ones, then you root out the bosses and don't look for that one. So it may well be quite subjective and have nothing to do with anything except the fact that there are always more people.

Weisskopf: When did the exponential rise start?

Morrison: With Newton.

In Unison: [Loud laughter.]

Telegdi: May I ask a question? Something has always interested me a great deal even before I went to the University of Chicago. You always see, not historical, but pseudo-historical, reviews which say that, well, Chadwick discovered the neutron, but there are other things, there's an address by Ruther-fore in 1920—I think it is printed in the Royal Society *[Proceedings]*, I think I read it—where there are some mentions made, and then there is always a reference to a professor of chemistry at my university, Harkins, and he had a number of quite distinguished students who became physicists. What was his role when you started your work?

Bethe: None.

Telegdi: Was it just numerology? How did he reach these conclusions?

Bethe: I don't know. It was numerology and nobody took it seriously.

Telegdi: Maybe they should have.

Bethe: No.

Weisskopf: Was he less serious than Elsasser?

Bethe: Much less serious.

Holton: There is a whole section of papers in the Ithaca Conference, of one of the International Congresses of the History of Science, devoted to the discovery of the neutron, to which Professor Segrè contributed, Purcell, Chadwick, and others, and you will find Harkins discussed in it to some degree. It is quite an interesting set. I had my eye on Chadwick for precisely the point that you raised here. There is the kind of thing which can only happen at a conference today and never gets into print in other form, that has to do with the attitude expressed by Chadwick in his personal memoir [at Ithaca]." I shall mention some of the more respectable of these attempts. There were others which were so desperate, so far-fetched as to belong to the days of alchemy."[24] He speaks about the search for the neutron. I wonder whether we can talk about the days of alchemy in the work of some of the people here which did not get into print—the desperate attempts that failed.

Morrison: There's one just so closely related to this that is not very important at all, and I don't wish to regard it as such—it is going from the ridiculous to the sublime. As a matter of fact, most of these things are just ridiculous, but once in a while, they are sublime. I suppose that when Rutherford, in 1920, used a discharge tube, in order to try to make a close combination of the electron and proton . . . . that's what he writes about . . . .

Goldhaber: This was later. '20 was a pure prediction. There was a fellow at the Lab—J.K. Roberts—who tried the discharge. That is published.[25]

Morrison: Yes, I know it. In the fifties, I have to say, that owing to the persistence of a young student, the same thing was tried, with better neutron

24. J. Chadwick, "Some Personal Notes on the Search for the Neutron," *op. cit.*
25. J.K. Roberts, "Relation between the Evolution of Heat and the Supply of Energy during the Passage of an Electric Discharge through Hydrogen," *Proc. Roy. Soc., A*, *102*, 72 (1922).

detectors and so on, because he maintained that it would still work, and with numerology, had many uses of the number 137 and 274 to make us not be able to destroy him within his own framework. So we allowed him to do this experiment. And I remember very well, Cocconi lent him the neutron detectors and the large file which he then used.

Weiner: I had an opportunity to discuss with Chadwick some of these "alchemy" experiments he referred to in his Ithaca paper. And the point is that it wasn't only for the neutron; there were many other things. It was a period of desperation, in the twenties, and they were ready in that particular laboratory to try many things. There was not very much to lose.

Cohen: That, by the way, I think is the first time in history in which a member of the Cavendish Laboratory used the word "alchemy" in a perjorative sense.

Goldhaber: You see Rutherford's first prediction really was rather deep because he didn't just say it would be nice if there were something like a neutron, but first he used the analogy—he had made one mistake, one of his famous experimental mistakes. He had found what he thought was helium-3 coming out of nuclei—the so-called long-range alpha particles. He did an e/m measurement which was incorrect. And while he was thinking that he had seen it, he was speculating what is helium-3. And he said, well, that is 3 protons and 1 electron. Then there ought to be something like 2 protons and 1 electron—we would call that heavy hydrogen. And then there would be 1 proton and 1 electron—we would call that the neutron.

But then came what I consider the deep remark: he said, "and come to think of it, how would the elements ever have been made unless there was something which could go through the Coulomb field?" This sentence is right in that prediction, and this is, I think, why it lived on.

Morrison: And he talks about the things having enormous penetrability.

Goldhaber: That's right. He says, "the neutron would just escape until it collides with the nucleus." He says all this in this one speculation—this is in the

Bakerian Lecture of 1920—and in these lectures he allowed himself speculations. He would not put speculations into papers, but in the lectures he would. And, you see, speculation was looked down on in the Cavendish. I lost out considerably because I agreed with that taste, not to speculate, because at the end of our photo[26] paper I had a paragraph on slow neutrons which Chadwick crossed through. I still have the manuscript. And the reason he crossed it through, which I quite agreed with, was sheer speculation. The reason for that speculation was that at that time Lea had that effect—seeing gamma rays with fast neutrons on paraffin—and we calculated the cross-section for that as the inverse of the photo-disintegration and his effect was a million times too large. So my last paragraph said that, well, this would be a million times too large if the neutrons remained fast, but maybe they are first slowed down in the paraffin. And Chadwick said this is sheer speculation. This was early '34, which I quite agreed with. My taste was exactly the same so I was quite glad to take it out. But I agree now this was wrong. In present thinking you would have written a paper just on that.

Morrison: Which is also wrong.

Goldhaber: Which is also wrong. But this is a little side history.

Segrè: That paper of Lea's was mentioned in the first slow neutron paper, in the discovery of slow neutrons.[27] We mentioned Lea because he was somebody known.

Goldhaber: Sure, but Lea did publish a paper later where he credits this speculation with explaining his effect.[28]

Wheeler: This comment of yours, Maurice, I think is very interesting because

26. J. Chadwick and M. Goldhaber, "A 'Nuclear Photo Effect': Disintegration of the Diplon by Gamma Rays," *Nature, 134,* 237-8 (1934).
27. E. Fermi, E. Amaldi, B. Pontecorvo, F. Rasetti and E. Segrè, "Azione di sostanze idrogenate sulla radioattività provocate da neutroni - I," *Ric. Scientifica 5* (2), 282 (1934). The reference to Lea is: "D. E. Lea, *Nature 133,* 24 (1934)."
28. D. E. Lea, "Secondary Gamma Rays Excited by the Passage of Neutrons through Matter," *Proc. Roy. Soc., A, 150,* 654 (1935).

it comes back to the point that was raised earlier about the extent to which people were individuals and the extent to which they were members of a group. This idea that Chadwick crossed out a section of a paper, as Bohr would cross out a section in a paper, or Breit cross out a section, or Oppenheimer cross out a section in somebody's paper—is there anything we know of that is more easily quantified and measured than this as a measure of the existence of groups in the world of physics?

Weisskopf: Did you make your Ph.D. thesis of this?

Goldhaber: It was my Ph.D. thesis, but the point was I think it was really partly the spirit of the time and especially the spirit of the Cavendish that you do not publish speculations.

Wheeler: Yes, but I am speaking now of another factor, that is, the existence of groups of people who have a certain standard of excellence.

Weisskopf: Johnny, I can report I was much upset in '36 when I made the terrible crime of sending a paper—it was the evaporation paper[29]—to the *Physical Review* without showing it to Bohr who happened to be in America at that time, and I got the only angry letter from Bohr that I ever received.

Wheeler: So this is another way of saying that the group has a certain standard that it insists on keeping.

Goldhaber: Yes, I think this is very important. Don't you feel it still exists, only it is more diffused because nobody has time to read all the papers? I should read all the papers which come out of Brookhaven for a standard.

Weiner: I just want to go back to the neutron thing for a minute. I looked in my notes from the Rutherford correspondence and I have since discovered that this letter is also included in Eve's biography of Rutherford.[30] It was

29. V. Weisskopf, "Statistics and Nuclear Reactions," *Phys. Rev., 52,* 295 (August 1937).
30. J. Chadwick to E. Rutherford in A. S. Eve, *Rutherford; being the life and letters of the Rt Hon. Lord Rutherford, O.M.,* 301 (Cambridge, England: The University Press, 1939).

apparently written in 1924 by Chadwick to Rutherford. Chadwick is on vacation, but he was discussing physics, and it's interesting. The only time, of course, that Chadwick and Rutherford exchanged letters was when they were away, when Chadwick was interned in Germany, or when one of them was on vacation, so it's those few times that we have this kind of connection.

In this regard, he says: "I think we shall have to make a real search for the neutron. I believe I have a scheme which may just work" . . . .

Weisskopf: From who to whom is this? I'm sorry.

Weiner: From Chadwick to Rutherford.

Goldhaber: At what time?

Weiner: About 1924. The letter is not dated but it falls into this category. Eve identified it in that period too.

"I think we shall have to make a real search for the neutron. I believe I have a scheme that may just work but I must consult Aston first." And this was the pattern throughout the period. It was consistent search. You have the feeling that the prospectus, the proclamation, was Rutherford's, but the carrying out of the program and the one who believed in it was Chadwick.

Goldhaber: Yes, he pursued it for a long time.

Weiner: And he was obsessed with this, not because it was just a bizarre thing, but because he thought that this perhaps would be the thing to open up the whole field. Which brings me to the next round of questions, or discussion, on the challenge to theory prior to the discovery of the neutron—the problem of electrons in the nucleus.

I am always a bit suspicious when people talk about a crises. It is always in retrospect a thing seems to be a crisis but I don't rule out the possibility that there was a crisis then, and I'd like to know: did people prior to the discovery of the neutron feel that there was really a crisis, or was it just the question of another problem?

Weisskopf: Well, what's the difference? It is a very fundamental problem.

Goldhaber: I don't know. My own feeling, though it is a little from a later time, is that at that time people just considered nuclear physics still something vague, where you don't get excited because every day something surprising happens. And crazy ideas could be suggested and this comes without a feeling of crisis.

Bethe: I spent the fall of 1930 at the Cavendish Laboratory and, of course, was surrounded by nuclear physics but kept strictly away from it. I heard of it in seminars and it seemed all totally uncooked and totally vague, and anything could happen. You couldn't get hold of it. The interesting subjects were quantum electrodynamics, scattering problems at high energies, and so on, that you could get hold of.

Weisskopf: But Hans, I was at that time at Leipzig with Heisenberg, and I had the impression that the problem of electrons in the nucleus excited the quantum electrodynamicists.

Heisenberg was terribly worried about it. There is a story about Heisenberg's worries: We were sitting in a coffee house facing the entrance of a swimming pool, and we were talking about the electrons in the nucleus. Heisenberg said, "Look, I see people going in the door and out of the door, all dressed. Do I conclude from this that they swim dressed?" This shows that the problem was a constant topic of discussion.

Segrè: I'll tell you one story about that time.

Weisskopf: You were there at Leipzig.

Segrè: No, I wasn't. But when there were the alternating intensities in nitrogen, the Raman effect work by Rasetti really clinched the argument that it had to be an even number of particles. Heitler had already said it, but the spectra were not so clear. I mean it was really this question of the alternating intensities. Well, Rasetti published it,[31] and somebody from Göttingen wrote him a letter to ask him whether the picture he had published was retouched.

Bethe: That is explainable because just before this, Rupp had retouched pictures of electron diffraction rings.

Segrè: And we kept on telling Rasetti all the time, "Where do you buy your brushes?" It was that bad.

Weisskopf: That also shows that people were very excited.

In Unison: They sure were.

Bethe: I'm sorry, I stand corrected on these electrons.

Caws: May I raise a question about this business of excitement. I was very interested by what Professor Goldhaber said earlier, that every day something surprising happens, therefore it is not interesting, therefore it is not exciting.

Goldhaber: No, no, I said the very opposite but perhaps poorly. It was exciting, but I could see that a man like Hans would say, well, since they change their mind everyday about the phenomena, why should I spend my important time on this? Not as rationally as this, but he must have come to this intuitive conclusion that the time was not ripe to enter with fundamental ideas in this field. There are many fields which are always on this border where you feel it is ready for an attack.

Caws: But you said that you would now not agree with Chadwick in striking out the purely speculative.

Goldhaber: No, I only say, the times have changed. I would say nowadays one would write a whole letter or paper on just this speculation. It would have reached that stage in the development. Whereas, at that time, even as a last

31. F. Rasetti, "Raman Effect in Diatomic Gases," *Proc. Nat. Acad. Sci., 15,* 234-7 (March 1929) and 515-9 (June 1929).

paragraph in an experimental paper, it was considered not good taste; whereas in a lecture, Rutherford was giving you speculations.

Caws: Yes, I understand, but what accounts for the change of style?

Goldhaber: Well, I think it is partly the pressure. I have also the feeling—there has been quite a bit of discussion on something which I often say in public lectures, and maybe it is worth saying here—I have the feeling that anyone brought up in the twenties, and this I am saying in retrospect to some extent, was brought up with the idea that there are protons and electrons. To think of a new particle seemed so radical that Pauli, who really was not the most timid of men, was extremely cautious when he suggested the neutrino, very apologetic. And it is a sign of progress how people now suggest a particle at the drop of a hat. Times have really changed.

Morrison: The actual increase in the number of particles is not great.

Segrè: There is something which is ununderstandable to me. Look at the first paper of Heisenberg, where he makes the nucleus composed of neutron and proton, but then he says, the neutron may well be made out of an electron and a proton, which, of course, if this were true the whole thing collapses. But it is written there, black on white. You can read it three times.

Weisskopf: But I don't agree. He probably thought that the neutron is a subunit, like quarks today, that neutrons and protons and electrons are somehow bound extremely strongly, and in the mechanism . . .

Segrè: But then the whole statistics, the whole spin, and everything else . . .

Weisskopf: True enough, he certainly didn't face the statistics.

Bromberg: No, but he took both positions in that paper. He said the neutron is both compound and elementary, that is, he says these two things at separate points, but basically he felt the neutron was still a mystery. Well, Mr. Wheeler was pointing out his whole attitude on that to me at breakfast time.

Wheeler: I was commenting on how Heisenberg's approach on these things was an empirical approach, how he had been upset about pair theory giving rise to particles and observed one had no evidence of this other particle. He was not willing to go as far as Dirac and change the mass of the positive electron to make it agree with the proton. So he had written this paper which tried to get rid of these negative energy states by introducing a lattice structure of states, only to discover that Carl Anderson had in the meantime discovered the positive electron.

Bromberg: But in those series of papers, in those three papers, he entertains both hypotheses that the neutron is elementary and that it is composite, and he uses the elementary hypothesis to discuss spin.

Wheeler: If I'm not mistaken, in one of those papers he even gives arguments why there can't be electrons in the nucleus just as in this earlier one of this series of papers that you cite he gave the argument or proposition that the electron was in the nucleus even though in a hidden form in the neutron. So he played it both ways.

Weisskopf: We should not be so much surprised about playing it both ways. For example, I mean, we seem to be surprised that Heisenberg didn't worry about the statistics argument. Today, the quark theorists are not worried about the fact that the wave function is symmetric and that the statistics don't work.

Segrè: At that time, at least in my recollection, people were much more struck and impressed by the statistics argument, at least in Rome, than by the argument of being contained or not contained, and so on. At least so it was in Rome because statistics was something strong in Rome. That was a yes or no answer; the other one maybe there is something there.

Uhlenbeck: Also the spin for nitrogen was one, that was also wrong.

Weiner: Henry Small has a 1930 Heisenberg letter that may shed some light on part of the discussion.

Small: This is written in German, so be careful when I translate. This is

Heisenberg to Bohr in 1930 and he is discussing his work in Leipzig. He says: "Here in Leipzig everything is as usual, my daily life and my work appear somewhat gray on gray. On relativistic quantum theory I attempt to think but I find until now no formal handle on it. Perhaps one must wait first for the complete development of nuclear physics until one here can go on."[32]

This is in 1930, and so somehow he was connecting the problems in the relativistic quantum mechanics with the problem of the electron in the nucleus.

Morrison: It's like the showers. Something is wrong with electrons at relativistic energies, maybe they interact there even though they don't show a shift in the normal spectra. You could have that point of view—something changes.

Small: If I could go on to a letter in 1932 which was written to Bohr just after Heisenberg wrote his first paper on nuclear physics. He has this sentence: "The basic idea of the paper is to blame all the main difficulties on the neutron."[33] And so this implies, I think, that the question really was how you deal with relativistic particles in the nucleus.

Telegdi: What was the German word used for "blame?"

Small: "Die Grundidee ist: alle prinzipiellen Schwierigkeiten auf das Neutron abzuschieben und im Kern Quantenmechanik zu treiben."

[General Laughter at word "abzuschieben]

Small: How would you translate it?

In Unison: That's right.

32. W. Heisenberg to N. Bohr, September 18, 1930. Bohr Papers, Niels Bohr Institute, Copenhagen. Microfilms of these letters have been made in connection with the American Physical Society-American Philosophical Society project "Sources for History of Quantum Physics" and are deposited at the American Philosophical Society, Philadelphia, University of California-Berkeley, and AIP Niels Bohr Library, New York.
33. W. Heisenberg to N. Bohr, June 20, 1932. Bohr Papers.

Holton: To shift over, Professor Bethe made a remark *sotto voce* to the effect that there were just more physicists at that time so that there was more efflorescence of ideas at the time. Do you remember, at that point the conversation was about why there was so much more speculation everywhere, and he said there were just that many more physicists.

Bethe: No, now. I meant there are now so many physicists and there are not that many fully developed ideas, so the physicists have to publish speculations in the absence of ideas.

Holton: The law of the conservation of ideas.

Goldhaber: The older generation of physicists is still against speculations. I'm told that when Lee and Yang had their famous idea, Rabi said, "Why do you publish it? Do the experiment."

Telegdi: I would like to ask you [Bethe] to make a comment about this 1930 visit to Cambridge when you felt that nuclear physics was too vague. But I think the converse was also true that you had plenty of things which were exciting and which could be done besides. That was one reason to reject it: you could do precise things which were exciting. Is that so?

Bethe: Yes.

Holton: Was there a feeling at the time that great centers are coming about? You look back and you see that there is Copenhagen and Berkeley, Caltech, Rome, the Cavendish Laboratory—each of them in a flourishing state. Compared to what happened some thirty years before, clearly it is quite a different affair. This differs from the early days of relativity. It didn't come out of large centers. It came here, there, wherever there happened to be someone who read it in the journals. Now, the professionalization is quite different. There are these great centers, and that is one of the reasons, of course, that there are students. I now see the way to put these things together. Where there are lots of students they obviously have one teacher near them. But previously, in the . . .

Weisskopf: When do you put this turning point?

Holton: It is not a turning point, but rather that there are now a large number of identifiable centers.

Weisskopf: I think there were always centers.

Holton: Not really, no. I think Copenhagen is 1921, when Bohr is put in charge. Berkeley, clearly.

Weisskopf: Sommerfeld.

Weiner: Göttingen.

Holton: Göttingen. Quite a few of these centers that are here identifiable are now working on the same kinds of problems. There are experimental facilities associated with quite a few of them, which of themselves require a larger staff. I wonder whether one can here think about the difference in professionalization when there are a large number, a critical number perhaps, of different centers, each of them having theorists and experimentalists—not in all cases, but in most cases—nearby. Does this raise some different image about what physics ought to be like at the time?

Weisskopf: I don't see with what one should compare it, because this is so since 1900.

Holton: Not quite, since the twenties, I would say.

Weisskopf: No, no, the Cavendish . . .

Holton: But that isn't new. Sure, there was also the Jefferson Lab since 1884, but it is a different style. In the Jefferson Laboratory, there are 2 Ph.D.s per year.

Telegdi: The Cavendish is much older than this.

Holton: Sure, but how many are there? That is the point. When you have the

Cavendish by itself, you have Göttingen, you have a very few working on different kinds of problems. Then, in the thirties, I'm playing with the hypothesis that not only are there more centers, but they are differently structured. They have a critical mass of people, both students and experimentalists and theorists. They are much more in contact with each other. They begin to have a feeling of physics being done in a different manner, and my model to compare it with—which is the study that I have worried about most—is relativity, for example, where the professionalization is entirely different.

Bethe: But if you talk about experiments, I think it goes back much farther. Take the discovery of the electron and of X-rays and the properties of the electron—they were done at different places, mostly England and Germany, from 1880 some to 1910...

Segrè: In France, don't forget.

Bethe: Yes, very much in France—by different people, in laboratories, each of which had quite a number of students, and of professors dictating what was to be done. But the laboratories certainly had relations to each other.

Shapere: The same thing is true in low temperature physics.

Bethe: Perhaps the main change is the connection between theory and experiment, and relativity theory, I suppose, was at a time when there was no very close connection. But I think the increasing connection between theory and experiment is perhaps the new thing that comes from 1920 on.

Segrè: That is a German thing.

Bethe: And Bohr, Bohr's connection with Rutherford.

Weiner: I wouldn't say that it was a national thing because within Germany you find different centers. For example, Professor Bethe told me about his experiences in Tübingen where there was not a close connection between experiment and theory. I think in our discussion we found that whether you had this close connection or not depended on where you were in Germany.

Weisskopf: In Göttingen, there was a close connection.

Bethe: In Munich, there was a fairly close connection.

Weiner: But it is not a national style. It happens to be a personal and an institutional style.

Bethe: I think at the same time in Berlin there was not a very close connection.

Segrè: But the professors of theoretical physics were all in Germany. There were no chairs of theoretical physics except for maybe Bohr. It was very different in spirit.

Bethe: That's right, but there were Bohr and Rutherford who had very close ties.

Morrison: And Thomson too, and all three were both theorists and experimenters.

Wheeler: E.J. Williams told me how he came to do theoretical work because Rutherford had this rule that the laboratory had to be closed up in the evening so he couldn't go back to the experiments that he was trying to do, and therefore he got going on calculations.

Goldhaber: Chadwick was most proud in discovering the neutron that he had to climb over the wall in the evening to get into his lab because it was all locked up. That was his proudest achievement.

Weiner: The feeling apparently at the Cavendish was that you need time to think and that you had better things to do in the evening than work in the laboratory.

Telegdi: Maybe they were right.

Weiner: And Chadwick in particular told the story of—I forget the name of the man who was the time-keeper there, essentially the man with the keys to lock up, who would come to the bench as people were working and just stand there with the keys dangling, letting everyone know that science must come to a stop at that moment. And this is the way it went. The place was closed, and it was considered bad taste, bad form, bad science, to work late.

Wheeler: I discovered that the worst thing you could do in England in writing a letter of recommendation for somebody is to say he works hard.

Morrison: That is just a gentleman's tradition. It isn't true any more.

Goldhaber: It isn't true any more. Now the Cavendish is open all day and night.

Morrison: Rutherford did not found the Cavendish. He did not make it so great. Maxwell, Rayleigh, Thomson...

Segrè: It wasn't just a speciality of the Cavendish to be closed in the evening. I remember Zeeman's laboratory was closed in the evening. I never remember anybody going in Rome to the laboratory in the evening. We never had to.

Holton: No, the paper of October 1934[34] was very exceptionally written in the evening.

Segrè: It was written, but it was not written in the laboratory, it was written in a house.

Uhlenbeck: The lab of Kamerlingh-Onnes in Leyden was always closed in the evening.

Telegdi: I want to ask you something. I have always thought myself that Kamerlingh-Onnes was really the inventor of what we call big physics today. He was probably the first man, even before Cockcroft and Walton, who brought

34. E. Fermi, E. Amaldi, B. Pontecorvo, F. Rasetti and E. Segrè, "Azione di sostanze idrogenate sulla radioattività provocata da neutroni I, *Ric. Scientifica* 5 (2), 282-283 (1934).

heavy, reasonably expensive engineering equipment to physics labs. Is this a clearly recognized fact?

Uhlenbeck: Yes, he was one of the first money-getters.

Morrison: Well, what about Dewar?

Uhlenbeck: The first low-temperature equipment of Dewar's was really huge. It was his whole laboratory.

Wheeler: Kelvin was a big operator, was he not, in the sense of the Atlantic cable? That's a fairly big operation.

Goldhaber: That was engineering.

Segrè: Edison was a big operator. He had an industry.

Telegdi: But then he was not a professor of physics, either.

Holton: Large equipment looked different in different years. In the Jefferson Laboratory in the period 1890-1910 people came to work with the big battery of 100,000 volts.

Morrison: Yes, Davy had a great battery.

Weiner: I would like to go from big things to small things again—particles and neutrinos in particular, and the theory of beta decay in the early thirties. I think Henry Small has some statements here that are interesting and I think this raises in a different way some of the issues that we have already touched on. We have about 35 minutes before we break for lunch and perhaps we can get into it during this period.

Small: First I would like to quote Professor Bethe from his interview:"Fermi's theory of beta decay in principle I think was accepted immediately because it was obviously a sensible explanation of the phenomena." I don't really want to challenge that statement or interpretation, but I think there were differences in the way that theory of Fermi was received. In particular, I think Chadwick expressed the view that the neutrino would not be easily introduced or explained in the electron or electromagnetic field theory. I'm paraphrasing here from a paper that he gave in London in 1934.[35] I know that Bohr did not receive the Fermi theory with open arms. So I wonder if this really was true that the neutrino idea really did catch on right after the Fermi theory.

Weisskopf: What is the basis of this statement about Niels Bohr?

Weiner: Shall we read the Gamow letter which is relevant?

Small: I think the Bohr position is given, partially at any rate, in a letter to Ellis in 1933, and this is before Fermi's paper. If I can quote here: "After my return to Copenhagen I have great pleasure in learning about the work of Doctor Guido Beck on the beta ray spectra of which he has also written to you. I am very much impressed by the progress as regards the explanation of the empirical effects he has achieved."[36]

Weisskopf: What did he do?

Small: Guido Beck's theory of beta decay did not use the neutrino.

Telegdi: His theory was that virtual pairs were created. There was a tremendous controversy about this.

Small: To continue: "But the final judgment about the correctness of his views will, of course, depend on the extent to which the consequences of the theory will be borne out through the continuation of experiments."

Weisskopf: That's a polite way of saying "It's nonsense." I know Niels Bohr.

35. J. Chadwick and N. Feather, "Nuclear Transformations Produced by Alpha Particles and Neutrons," *Proceedings of International Conference on Physics, London 1934,* 95 (Cambridge University Press, 1935).
36. N. Bohr to C.D. Ellis, August 30, 1933. Bohr Papers.

Weiner: There is another letter—this is Gamow to Goudsmit—describing conversations with Bohr on this subject.

Small: And this is in 1934 after the Fermi theory. "Bohr, on the other hand, well you know that he absolutely does not like this chargeless massless little thing, thinks that continuous beta structure is compensated by the emission of gravitational waves which play the role of neutrino but are much more physical things."[37]

Goldhaber: More weightless.

Segrè: Fermi thought that the neutrinos were the quanta of the gravitational field. He really did.

Small: We do have evidence that in 1936 Bohr did accept the neutrino because there is a definite statement in a letter that he did accept the idea.

Bethe: In 1936, yes. Well, I remember there was still a hot controversy about energy conservation and that we were eagerly looking for determinations of the maximum energy and conservation of energy vs statistical energy conservation which Bohr had plugged in '31.

Segrè: At what time were you looking for this?

Bethe: I'm now talking about '35.

Segrè: One year after Fermi.

Goldhaber: But at that time was the Ellis-Mott paper which had very good proof.

Wheeler: Yes, that was the paper in October 1934. That was one of the most

37. G. Gamow to S. Goudsmit, March 8, 1934. Goudsmit Scientific Correspondence. Microfilm copies are deposited in Niels Bohr Library, New York; the original letters are in the possession of S. Goudsmit.

interesting things at that Nuclear Physics Conference of October 1934, this Ellis paper showing the upper limits would give energy conservation.[38]

Telegdi: I don't recall the details but all I can tell you is that it is worth reading these things. First of all, at the Conference that Professor Wheeler is referring to, there was a report by Beck on his work. There was a tremendous controversy in which Fermi, I think, only participated in a very weak way, saying well, what I am doing is the physical theory, what you are doing is a pipedream. I think what Beck was doing was some sort of a quanta series, which however had this Bohr type of non-conservation in it. And there was quite a controversy and I think it was not at all obvious except to a few people that Fermi had a complete theory and Beck just had a mathematical version of a theory without real conservation. Beck has a report. Of course, he takes his own point of view, and then he says at the end, the distinction is yet to be made.

Weisskopf: There is a certain distinct difference in the styles of Fermi and Bohr; somehow the paper of Fermi may not have impressed Bohr because of its style, the style of approach.[39] The whole idea that an electron and a neutrino were emitted together, of course, was there before Fermi's paper.

Uhlenbeck: That was Pauli.

Weisskopf: Yes, we had Pauli way back. It was discussed all the time, and now Fermi did put this in a very lucid mathematical form. I believe that Bohr did not regard this paper as a great step forward.

Telegdi: But still Fermi's paper had something which nobody else had. It had the $F\tau$ value.

38. C.D. Ellis and N.F. Mott, "Internal Conversion of Gamma Rays and Nuclear Level Systems of Th B and C." *Proc. Roy. Soc., 139,* 369 (1933); C.D. Ellis, "The Beta-Ray Type of Radioactive Disintegration," *Proceedings of International Conference on Physics, London 1934,* 43 (Cambridge University Press, 1935).
39. E. Fermi, "Tentativo di una teoria dell'emissione dei raggi 'beta' " *Ric. Scientifica, 4* (2), 491-495 (1933); also, "Tentativo di una teoria dei raggi beta," *Nuovo Cimento, II,* 1-19 (1934); "Versuch einer Theorie der Beta-Strahlen, I," *Z. Physik, 88,* 161-171 (1934).

Morrison: But there was much doubt about that. For example, did you not put in gradient wave functions? The shapes were wrong . .

Telegdi: That's true because the experiments were incorrect. Nevertheless . . .

Bethe: But those are variants just like Wigner and Majorana.

Morrison: But in a sense Fermi's paper is only an expression of Pauli's idea.

Telegdi: That is not true because if you read that paper it has already one thing in it that to my knowledge—I once worked on a short history of weak interactions, and I read those papers. The other papers had nothing quantitative to say, and Fermi's first paper had something quantitative to say. He said, well, I can connect the lifetimes of several beta decay elements . . . .

Uhlenbeck: And it had the influence of nuclear fields. That was really Fermi's idea that you could do that. Still, I agree with the thing that it was, so-to-say, a formulation using the accepted formalism.

Telegdi: What I am trying to say is that there are two things: first, having understood field quantization he managed to explain the emission of particles in a mathematically satisfactory way—that's very nice—but that somebody else could have done, like Pauli. But that was in there. In fact, he had to do a lot of hard mathematics because the only beta decaying elements in those days were very heavy, so he had to do all the complicated things with Coulomb wave functions and so forth, and then he made these $F\tau$ values, and these other rival theories just had only mumblings. This is typical of Fermi; there are numbers in the paper.

It's a very strong point that in the first paper there are $F\tau$ values and comparisons, and saying, then this must be so.

Goldhaber: What was Perrin's contribution to the beta decay theory? You know he made this one remark at the Solvay Congress in '33 that this positron spectrum looks like a beta decay, and then he had a rough formulation that the neutral particle must be lighter than the electron because of this spectrum.

And he had a little bit qualitatively of what Fermi developed. I wonder whether this affected Fermi.

Telegdi: I think it was after Fermi.

Goldhaber: No, it was before Fermi. It was at least a few months before.

Telegdi: There are two papers by Fermi, you know. There's one in Italian and one in German.

Segrè: They are two weeks apart. Fermi knew of Perrin's idea when he wrote the beta decay [paper]. He may well have been influenced by this. He certainly knew of Perrin because he had heard Perrin at the Solvay Congress.

Goldhaber: Maybe he quotes it. I just don't remember.[40]

Telegdi: The only person he quotes is Hulme because of the Coulomb wave function. He reports that.[41]

Small: If we assume that there was some skepticism with regard to the existence of the neutrino, could we ascribe this to the state of the experiments at the time of the beta spectra, or is this a resistance to take up a new particle?

Bethe: I think it was the latter, and especially one that you couldn't observe.

Shapere: Was it because it was new or because it was massless?

Bethe: It was because it was new and unobservable, I think, more than massless. After all, there was the photon.

40. In the above-cited *Z. Physik* article, Fermi has the following footnote: "In einer kürzlich erschienenen Notiz kommt F. Perrin, *C. R. 197*, 1625, 1933, mit qualitativen Überlegungen zu demselben Schluss."
41. Another Fermi footnote from the same article: "R. H. Hulme, *Proc. Roy. Soc., A, 133,* 381 (1933)."

Segrè: At the end of the beta spectrum, they could say, well, if the mass is very small, we wouldn't have seen it.

Telegdi: As the paper stands, there is no assumption of masslessness. He says it's a small mass and discusses again how experimentally you would tell. It's by the end points.

Weisskopf: I think it was the *ad hoc* nature of the neutrino.

Morrison: Also the unabsorbable nature so that it raised a whole new box of troubles.

Uhlenbeck: I do think though that that paper remarkably stimulated experiments because for the first time people began to measure shapes. Just from my personal memory, when Konopinski and I tried to compare it with the shape, it didn't fit with anyone, and then by putting a derivative in, it suddenly fitted. So we thought that had to be done. And I remember Robert Oppenheimer was quite impressed by it because Kurie was then measuring shapes and it fitted always the KU plot in those days. Now all these experiments were wrong. And then it took, I think, five years.

When I came back from Holland in 1939, I was there in Ann Arbor, and there was suddenly Lawson, who had done, I think, indium for the first time. He said: "It's precisely the Fermi shape."

Weisskopf: Why were they wrong?

Morrison: Multiple scattering in the sources.

Telegdi: The sources were too thick.

Goldhaber: The only thin source was unfortunately not a Fermi type. It was Radium E. And every student was told by his professor: "You build a beta spectrograph, then to test it you use Radium E." Then by that time he was usually too old to do anything more. He always published just Radium E.

Small: If I could follow up my previous question: If it was because the neutrino was such an unusual new particle, then what did convince the people that this was the necessary explanation? In other words, was it the experiment that convinced them?

Goldhaber: I think there was the strong Pauli argument that energy is not conserved without it and statistics and spin counted too, but that energy was not conserved was a tremendous argument in itself.

Morrison: I think that's why Bohr is such a bad authority on judging these papers because Bohr really had it in his mind that there was some profound problem with neutrinos and energy and so on, and didn't want to have it solved except in a mystical and a deep way. It was solved by Fermi in too elementary a way.

Small: I'd like to quote from a Bohr letter in 1941. He is writing Gamow and recollecting the time in which the beta problem was discussed. This is from Bohr to Gamow in '41 and Gamow had just sent him an article on stellar collapse. Bohr comments: "When reading your article my thoughts went often back to our discussions in Copenhagen, when you started on this fruitful line of work. I remember especially our talks about the B-puzzles and even if we then had no idea of all what should happen since, I feel, that those of us who tried to stick to fundaments may have quite a good conscience."[42]

Goldhaber: This is a great illustration of style, style as made by one man and by a stubborn man.

Segrè: Conservation of energy—I would call that fundamental. Now the people who were trying to stick to fundamentals in my opinion were Pauli and Fermi and the people working with neutrinos. Does he say that or does he say the opposite?

Telegdi: I have a letter of January 1957 from Pauli. He replied to my having

42. N. Bohr to G. Gamow, March 27, 1941. Bohr Papers.

sent him a certain number of preprints about parity breakdown, the Columbia experiments and other experiments. And in this letter he said he got my letter just two hours before delivering a lecture on the history of the neutrino, and he said he was never, or rarely, as upset in his life as having found out all these new things just a few hours before his lecture. And then he says in this letter: "Maybe Bohr was after all right when he insisted in 1932 that there is something fundamentally revolutionary in beta decay." I have the letter.

Morrison: And Bohr too wrote such a statement after this. There is still some mystery about the neutrino we have not yet perhaps understood.

Weisskopf: Bohr and Pauli said that things are not as glib and simple as Fermi makes them. That is the attitude.

Bethe: But this was really something that could not be foreseen.

Weisskopf: *Schadenfreude,* that things are not as good as all that in beta decay.

Telegdi: It was only Germans who had that feeling.

[General Laughter]

Holton: I wish that you would elaborate just a little more on this idea of style differences.

Weisskopf: It is a very difficult thing. Obviously there are styles, and obviously Bohr and Fermi are representatives of two very different styles. Johnny, you know more about this, tell them.

Wheeler: I would say that the difference is more in having a sense of wonder than in any other single thing.

Weisskopf: Yes, I believe that's a good way to say it.

Bethe: One example of Bohr's attitude was given at another Solvay Congress.

I think it was '58 or thereabouts, when Feynman gave a talk about quantum electrodynamics and how well everything fit. And Bohr said, "This is terrible that the problems are solved in one closed area, and then we cannot make any progress."

Morrison: The same thing he said at the Princeton meeting, when was that?

Weiner: 1946.

Morrison: The very same thing. He kept saying, you will not do this without bringing together other problems. You cannot solve it.

Wheeler: Always the point that if you had one problem you were in trouble, but if you have two, you've got hope, because you can always work them off against each other.

Morrison: And Fermi's was certainly just the opposite style. Fermi always spoke of success in what had been done and Bohr really always on what was not yet done and what the contradictions were.

Bethe: And Fermi wanted to solve the problems one by one.

Telegdi: I have never known Bohr so I cannot say. But Fermi, I think, not to speak of style, but of philosophy, was just very pragmatic. The theory is good because it explains the facts, and to hell with all philosophical sidelines. In quantum mechanics I never heard him discuss the mysteries of measurements—this whole religious attitude that quantum mechanics works because you compute things and they come out right, and what could be as strong a proof as experiment. It's a very pragmatic point of view.

Holton: May I follow this up by asking this? There is a feeling that physicists often were interested in philosophy or philosophy of science, that some read Reichenbach or knew Carnap or listened to Unity of Science meetings and the like. Did any of the physicists here feel that philosophy of science made much of a difference during the working out of the early story of nuclear physics itself? Did you feel a member of a philosophical tradition of Mach or someone else?

Weisskopf: It is very hard to say because if you are, as I was, under the influence of Bohr at that time, obviously the whole atmosphere was "philosophical" and directed towards fundamental questions. Bohr couldn't talk about nuclear physics for ten minutes without coming back to the uncertainty principle, justifiably or not.

Let me quote one mean remark of Pauli that George [Uhlenbeck] told me. Pauli said that Fermi is a "quantum engineer." There is something in this remark—it shows the difference of style. I am a great admirer of Fermi, but there is something in it.

I would only conditionally say "yes" to your question. It is hard to make a counter-experiment. In other words, I don't know whether nuclear physics would have progressed faster or less fast at Copenhagen, if the Copenhagen atmosphere would not have been full of complementarity.

Bethe: I think the compound nucleus was strongly influenced by complementarity. On the other hand, there were Breit and Wigner who were not so influenced, who found something which mathematically was equivalent although the philosophy was different.

Weisskopf: I am worried—since you're talking about it—about the equivalence of Breit-Wigner and the compound nucleus. I would actually like to make a case that this is just not so.

Breit-Wigner is one part of the compound nucleus, going further in that special direction, because it is a quantitative theory. But, for example, how could you ever get the statistical theory of nuclear reactions out of Breit-Wigner?[43]

The idea of Bohr—the way he talked about the nucleus—is different from Breit-Wigner.

Bethe: Yes.

43. G. Breit and E. Wigner, "The Capture of Slow Neutrons," *Phys. Rev., 49,* 519 (April 1936).

Weiner: Perhaps this is a good time to get started, for ten minutes, on this subject of the compound nucleus and the general development of models. I think maybe it is best to start with theories of the neutron resonances.

The obvious question suggested is the effect of the discovery of neutron resonances. We have pretty good documentation of the uproar in Copenhagen at the time—the discussions as a result of this, and the work that followed. Was this seen, again, as any crisis for existing ideas of the nucleus?

Weisskopf: Again, it depends on how you define crisis. The results of Bethe's paper and Fermi's discovery were certainly discussed in Copenhagen day and night at that time.

Goldhaber: It showed perhaps Bohr's style when he had the compound theory he came to Cambridge and gave a talk on it. I think it must have been the Fall of '36—would that be about the right timing?

Bethe: Too late.

Goldhaber: I mean after it was published he came over. It was published shortly before, wasn't it?

Weiner: Here is a letter from Rutherford to Chadwick, 17 February 1936. "We have just had a visit from Bohr who was full of his new qualitative theory of the nucleus. He considers that Heisenberg and Gamow and Co. have all got hold of the wrong end of the stick and that it is impossible to deal with the movements of an actual particle within a nucleus. He therefore considers that the space inside the potential barrier is a sort of 'mush' of which the vibrations can be qualitatively worked out on wave mechanics. One result of his theory is that there should be a number of very sharp levels with very small energy differences which serve to account for the peculiarities of neutron absorption."[44]

44. E. Rutherford to J. Chadwick, February 17, 1936. Rutherford Papers, Cambridge University Library. Microfilm copies of the Rutherford Scientific Correspondence are deposited at the AIP Niels Bohr Library, New York.

I might say this is a very interesting example. Here is Rutherford reporting, and you might comment, from your own experiences, how accurate he is in reporting what Niels Bohr was saying.

To continue: "He also thinks he can explain in a sensible way why some elements have such a big target area. He is publishing a note on his theory in *Nature* shortly. I myself am inclined rather to like his point of view and it certainly seems to help us in understanding some of the special properties of neutrons. He thinks he can explain in a natural way the levels in the radioactive bodies and he also considers that a nucleus can be excited to almost any extent without breaking up at once."

Bohr left in the morning. This was written in the afternoon. Chadwick replied a week later. "I was very interested to hear about Bohr's new views about nuclear structure. It seems to me to be coming back to our earlier ideas when we considered the nucleus as a whole and talked about the energy state of a nucleus without putting a special particle into a special place. Hulme was here a few days ago and told me a little about it, but I did not understand completely how the quantitative predictions appear. I shall have to wait until the note comes out in *Nature*."[45]

Weisskopf: It is most impressive how Rutherford understands physics.

Goldhaber: It must have been at that time that I heard Bohr's talk. I had forgotten the exact time. And I remember asking how about the small cross section of bismuth, which was already known to be small, and, of course, it was interesting that Bohr considered this just a minor little problem which he would not like to get absorbed in compared to the big problem of explaining nearly all the nuclei. And, of course, this was one of the future shell model nuclei.

Segrè: As far as I can remember, my reaction, and probably the reaction in Rome, was that this thing of a compound nucleus, nobody knew what it was.

45. J. Chadwick to E. Rutherford, February 25, 1936. Rutherford Papers.

It was sort of philosophy, not really a theory, but the Breit-Wigner formulae were really a thing—you go there and you measure, you find the energy at which it resonances, you find the width, you find the cross section as a function of energy; you can test.

Weisskopf: I don't understand this.

Bethe: But they are complementary.

Weisskopf: Yes, you must explain why are the widths so small, why are the level distances so small.

Bethe: Yes.

Segrè: I don't know. I just want to test it. The Breit-Wigner formula is something testable.

Telegdi: It doesn't give you anything about the spacing [of the levels].

Uhlenbeck: But there are other things testable too.

Morrison: It is just a parameterization of the experiment to show it is quantum mechanical. That of course was useful because the cross sections were so large you worry about them.

Segrè: The worry was why is capture so colossal and scattering so small. How can this be? Breit-Wigner told you. You measure it.

In Unison: No, no, no.

Goldhaber: Bohr told you.

Weisskopf: Breit-Wigner does not say that the radiation widths are much larger than the scattering widths; it only expresses the resonances in terms of the widths. The widths are parameters. They could have any value. I understood that Breit-Wigner and Bohr were two quite different things. One is a mathematical formulation, a quantum mechanical formulation of the phenomenon of resonance, saying nothing about the widths. But somehow Wigner and Breit do seem to have an idea behind it.

Wheeler: It was never clear what the idea was. There was this picture of two particles interacting but it wasn't this picture of a compound nucleus.

Bethe: There was the picture of two particles interacting and this would certainly increase the number of levels. And now you could say—and I vaguely remember that that's what I said to myself—that there are not two particles but there are 100 particles in silver. So you have many more particles interacting and therefore the spacing will be much smaller. And I think that is what Breit and Wigner must have had in mind. I remember I had discussions on this with Placzek on this very point: is Breit-Wigner the same or different? I claimed at that time, as far as I remember, that Breit-Wigner already had at least the kernel of the Bohr compound nucleus. They had something else too, the parameterization, but that it was not fundamentally different.

Weisskopf: Is it in the paper, the parameterization?

Telegdi: Is it not true that Breit-Wigner was a perturbative treatment?

Bethe: It is.

Telegdi: That is anti-compound nucleus because the compound nucleus says whatever hits the nucleus sticks and that can hardly be a perturbative process.

Morrison: But then there was a whole big thing which grew out of that, whether now that you've got to set up the potential scattering and resonances you can break off the perturbation theory. There are many learned papers—Peierls, and so on—trying to figure out how this works.

Telegdi: No, but that Peierls-Kapur paper[46] is much more like what Viki [Weisskopf] did. It was a boundary fitting condition, an anti-perturbation

46. P. L. Kapur and R. Peierls, "The Dispersion Formula for Nuclear Reactions," *Proc. Roy. Soc., A, 166,* 277 (1938).

theory. It seems to me that a man who does perturbation theory really doesn't believe much in the compound nucleus which is a paragon of strong interactions. Everything that comes in just sticks.

Morrison: That really isn't quite fair. It is like dynamics—you also derive a perturbation, it diverges, you cut off, and say this is good.

Telegdi: That's a different thing, though.

Morrison: It's an analogy.

Telegdi: No, no.

Weisskopf: Val, you have such a good memory, and you probably read the Breit-Wigner paper. Does it say anything about why for example radiation is so much larger than the scattering?

Telegdi: No, I haven't read the paper certainly for 10-12 years, and it may be 15, but as I recall, my feeling was that these people knew very well—Wigner, Weisskopf, and all these things—they tried to find a trick to drag in the well-known resonance formula of atomic physics into nuclei as hard as they could. They had this background of how you derive a Breit-Wigner shape in atomic physics and now let's try to drag it into nuclear physics by means which to me later seemed completely illicit because of the strong interactions.

Weisskopf: That's my impression too.

Wheeler: It's the right answer to the wrong question.

Telegdi: No, it's the correct answer to the right question but a wrong meaning.

Morrison: That's what I'm saying. You get an answer from perturbation theory which you hope is better because it fits the experiment.

[Adjourned for Lunch]

Scenes from the May 1969 Conference. (Photos by Heka.)

Above: Beginning in the foreground, Philip A. Morrison, Robert Serber, George E. Uhlenbeck, Lewis Slack, Maurice Goldhaber and Victor F. Weisskopf.

Below: Hans A. Bethe (left) and Charles Weiner.

Above: Maurice Goldhaber (left) and George E. Uhlenbeck
Below: Victor F. Weisskopf (left), John A. Wheeler and Martin J. Klein.

Session II. Postwar Nuclear Theory

Charles Weiner, Chairman

Weiner: Before we continue from where we stopped and then tackle postwar nuclear theory, I would like to play a tape of an excerpt from an English language broadcast made by Niels Bohr in 1938. (It was made by the Danish Broadcasting Co., and our copy came through the courtesy of the Atomic Energy Commission in May 1968.)

Bohr: In recent years, we have entered a new era in physical and chemical science, initiated by another of the great discoveries with which Rutherford's genius has presented us. Between 1900 and 1910 Rutherford succeeded in proving that the nucleus of the atom, the intransibility of which under all ordinary conditions is the very basis for the constancy of the elements, may under extreme circumstances afforded by the bombardment of high-speed particles itself undergo disintegrations. This modern revival of the old aims of the alchemists has, in the last few years, led to a succession of most startling discoveries regarding the ultimate constituents of matter. It has already found very important practical applications, especially through the discovery of artificial radioactivity by the daughter and son-in-law of the immortal Madame Curie.

Above all, this discovery has given us an entirely new means of studying the incessant exchange of matter within living organisms by means of the indicator methods originated by Professor Hevesy, the privilege of whose collaboration we enjoy in this Institute.

The intense world-wide interest in this young nuclear physics has given rise to great efforts in developing ever more powerful tools for this kind of research, ranging from the construction on a gigantic scale of electrostatic machines akin to those of quantum states to the design of electromagnetic accelerators of atomic particles on quite new principles. On a recent visit to America, I received an unforgettable impression of the success of these courageous efforts. It is a most pleasant duty on this occasion to express my special indebtedness to the most ingenious inventor of the cyclotron, Professor Lawrence of Berkeley, who has so generously put his precious experience at the disposal of this Institute, as well as of many other laboratories all over the world.

In building such complicated and costly apparatus, the usual aim of scientific investigation in practical industrial activity is felt particularly strongly on account of the technical perfection required as well as of the necessary financial support. In our case, this support has been most liberally contributed by the Rockefeller Foundation and by Danish foundations which likewise owe their sources to industries which again depend essentially on science.

In concluding this short address, I wish to give expression to the hope that the example of the fruitful cooperation of science independent of national borders and prejudices may stimulate efforts to contribute by peaceful collaboration also in other fields of human culture to the common progress of mankind.

Weiner: And following Bohr is a small excerpt from a recorded speech made by Rutherford at Göttingen on December 14, 1931.[47]

Rutherford: Now I hope I'm not tiring you too much. There's another theoretical point I would like to mention.

So far, we've spoken of the origin of the gamma rays. But now the question is: is there any connection—can we explain—the complicated system of gamma rays given, say, by Radium C. It consists of probably 50 or more gamma rays. Those energies are—some of them accurately, others roughly—known. What point of view can we take on that very difficult problem?

The bother is that a nucleus, as you know, is a very small thing and we know very little about it, and this is one of the methods, by trying to find these relations, that we can hope to find something about the nucleus.

Now, you remember . . . I have had the opinion for a long time—that's a personal conviction—that if we knew more about the nucleus, we'd find it was a much simpler thing than we suppose. These fundamental things, I think, have got to be fairly simple, but it's the non-fundamental things that are very complex usually. So we are hopeful . . . we must look for simplicity in the system first. And if the simplicity—we can't find it—well, we've got to look at something more complicated. I am always a believer in simplicity, being a simple person myself. (Laughter from Rutherford's audience)

Some discussion lost in switching tapes
from the Bohr and Rutherford talks to
responses of the Conference participants.

Morrison: This reminds me again that nuclear structure is already implicit in five or ten years of working with radioactivity . . . you really had spectroscopy.

Weisskopf: I don't understand.

47. Recording of Rutherford's December 14, 1931 talk at Göttingen deposited at AIP Niels Bohr Library, New York.

Morrison: He was saying there that it was already in the short-lived highly excited daughters of the natural series that the whole complexity of nuclear spectroscopy appears. And they knew that in the pre first World War period or just about the first World War. They didn't understand it but they had the beginning of the characteristic energies.

Uhlenbeck: It is a bit surprising that he thinks it's simple.

Morrison: Yes, well, he sort of hopes. He's a little bit nervous there, I felt, because he says if it's not simple, maybe we'll have to go to another . . . but he still hopes.

Weiner: Many things were to happen. That was December 1931. The following month was the beginning of the neutron, so it was just on the brink of that.

Weisskopf: It was a great year.

Weiner: We left off when we had begun to talk about the compound nucleus and the general question of models of the nucleus or pictures of the nucleus. I'm not clear whether, in fact, the word "model" was used at the time or whether this is a later philosophical invention, but I don't think that is really the point. It is like the word "crisis," which was not the issue.

The thing that seems to be of interest was the competition among various pictures or models of the nucleus, and the fact that, as John Wheeler remarked in earlier discussion, there were several co-existing, and apparently adequate for separate purposes, pictures of the nucleus. I wonder if we can talk about the development of some of these pictures over a period of time. A couple of things come up here—not only the compound nucleus but some of the reasons why the shell model in its early stages wouldn't do, and whether there were resistances to accepting it later on when it reappeared, and what the basis for these hold-outs was. I would now like to just open up the floor.

Wheeler: In a certain sense the first paper of Heisenberg's on nuclear constitution after the discovery of the neutron—well, it was more than one paper—

implicitly adopted the shell model, one could almost say, because there was the idea of potential, the idea of particles moving in that potential, and a wave function, and states for these particles. So one could really ask why didn't that get followed further.

Bethe: It did.

Weisskopf: Hans Bethe wrote a paper.[48]

Bethe: I think it did get followed and I don't think there is a contradiction between the idea of a potential between two particles and the compound nucleus. There is no contradiction between the Coulomb potential and a liquid. It seems to me that the two models, the shell model and the compound nucleus, were applied and were explaining different features. I heard Niels Bohr say at one time: "Well, the deuteron can be explained by a potential. With the triton, I'm not sure." He took an extreme point of view, but certainly when you come to something like a medium-weight nucleus, 20 or 30 or so, you have to use the compound model. But you were looking at different features. You were looking at highly excited states of complicated nuclei in the case of the compound nucleus, and you were looking at low states of scattering of very simple systems.

Weisskopf: I believe in the heydays of radical compound nucleus—that means in '37, '38, '39, in fact, through the war for some of us—we did not restrict ourselves to highly excited states. It was the general idea to take a heavy nucleus from iron on, that the whole spectroscopy is essentially statistical, and actually you did, and I did, level-density calculations which are supposed to be valid pretty low down.

Bethe: I guess so, although we applied it always to highly excited states.

Weisskopf: Well, where the neutron is absorbed, but that was only because we were interested in that region technically.

48. H. A. Bethe, "An Attempt to Calculate the Number of Energy Levels of a Heavy Nucleus," *Phys. Rev., 50,* 332-341 (August 1936).

Bethe: 8 MeV, though.

Morrison: Wasn't it true that we understood in those days about inelastic scattering in lead which was worse than other elements for tamping?

Weisskopf: . . . . the nuclear temperatures were of the order of 1 MeV and that means you already assume the levels up to 1 MeV should be somehow statistical.

Bethe: But it still points out, we knew that wasn't so. The first level in lead is over 2 MeV, and the first level in uranium was 40 kilovolts.

Goldhaber: That was not known at that time.

Bethe: It was known that it was very low.

Goldhaber: It was not known that it was that low.

Bethe: It was not known that it was that low, but it was known that it was maybe 200 kilovolt.

Goldhaber: Maybe it was secret, I don't know.

Bethe: I don't know whether we knew the first level. I have some . . .

Weisskopf: I would say, at least as far as I and my close friends are concerned, it was of course know shortly after the war and during the war from these things that you mention, and from other things, that not everything is all right with the compound nucleus. And then when the shell model came in 1949, there was a big shock, the first one for several years. It looked like a contradiction. How can we speak of single particle models and at the same time have these narrow resonances? And I think it took a number of years before this contradiction was resolved. After all, this is only a matter of time scale. If the particle runs long enough in a relatively free potential, then after some time the energy is dissipated and the fine structure of the levels becomes important. So it took some years, but then evidently, it came out that fundamentally there is no contradiction between the two models.

Bethe: I think I would say it somewhat differently, namely that it wasn't so much the contradiction but how could you understand that the shell model could be true at all.

Uhlenbeck: The short-range forces.

Weisskopf: Yes, of course, but then, I think came this idea . . . . I wrote some kind of qualitative paper in *Science*[49] that the Pauli principle . . . .

Telegdi: You wrote it in Zurich.

Segrè: That story of the Pauli principle was already in the first edition of Fermi's Notes, I think.[50]

Bethe: As a justification of orbits?

Segrè: Yes, that the Pauli principle prevented collisions. That must be in the first edition of Fermi's Notes.

Bethe: Which was when?

Morrison: 1949. That is the same time.

Weisskopf: Well then, I stole it from there.

Telegdi: I remember I was in Zurich when you wrote this paper and I read it.

Weisskopf: But it is quite true, I can still remember, this was the first shock. How could it be—with short-range forces?

49. V. Weisskopf, "Nuclear Models," *Science, 113,* 101 (1951).
50. E. Fermi, *Nuclear Physics.* A Course Given at the University of Chicago. Notes compiled by J. Orear, A. H. Rosenfeld and R. H. Schluter, 111, (University of Chicago Press, Revised Edition 1950).

Telegdi: I don't know whether you remember that at the Basel Conference there was some discussion Heisenberg brought up and said he could not understand why it did work but it certainly was amply demonstrated that the shell model worked.

Bethe: When was that?

Telegdi: 1949.[51]

Weisskopf: Why did we not at that time draw the analogy with the electron gas in metals?

Telegdi: You did. It was in that paper.

Weisskopf: I mean at the beginning.

Telegdi: I think because you were too happy about the successes of the vibrating drop and this and that and the other.

Wheeler: And after all it was an accident that the mean free path was so nearly comparable to nuclear dimensions. It was so easy to think either it was very short or very long, but anybody who thought that it was that same order would have been thought to be a dreamer of dreams. It is just unbelievable that two such different models should work so well, and it was only this accident that made it possible. Nobody had any right to anticipate that accident.

Small: Could I make one comment? Before lunch, a letter from Chadwick to Rutherford was read. One statement was that the compound nucleus appeared to be coming back to earlier ideas on the nucleus. And I'd like to put this to you: What ideas were they specifically coming back to?

Goldhaber: He explains in the next sentence.

51. "Diskussion zum Vortrage von Prof. L. Rosenfeld," *Helvetica Physica Acta,* in Basel, vom 5, bis 9, 225 (September 1949).

Bethe: Read the next sentence.

Small: "It seems to me to be coming back to our earlier ideas when we considered the nucleus as a whole and talked about the energy state of a nucleus without putting a special particle into a special place."

Uhlenbeck: They were so strongly coupled that you cannot speak about individual particles any more, I think that's probably what he means.

Small: This implies that there was some earlier version of the compound nucleus, perhaps, that was in some way related to Bohr's idea.

Goldhaber: Rutherford used to always write the reaction A+B into C first, and then C breaking up into D+E, before the compound nucleus, and this became physically meaningful only after the compound nucleus.

Segrè: Oh, did he do that?

Goldhaber: Yes, he did do that, and it was more than just a convenience. He really used to think of that. You can see this in some publication probably. I remember him always doing that, and I didn't particularly appreciate it until later.

Telegdi: But if you think of the experiments that were done in the Cavendish like bombarding lithium 7 with protons which had very distinct resonances, it's natural to . . .

Goldhaber: Yes, then he would put beryllium 8 in between, and then later . . .

Telegdi: The early resonance experiments responded to specific incoming energy which then would decay.

Bethe: Which were the first experiments of this kind?

Telegdi: Lithium 7 plus protons.

Goldhaber: But the first one was artificial—they had resonances earlier with alphas.

Telegdi: I'm speaking of the ones that were done in those years which I talked about at the Cavendish, and some at Caltech.

Bethe: Caltech, I remember.

Segrè: They were at the time hunting for resonances with alpha particles in Bothe's laboratory before the discovery of the neutron.

Bethe: I remember those.

Segrè: Remember Pose?

Bethe: They were then resonances in elastic scattering and that was fine. You had a potential between an alpha particle and the target nucleus, but I don't remember that this was applied to the reactions in which, let us say, a proton was produced.

Segrè: Look, I can't remember the dates, but I remember a sphere that had evaporated beryllium inside and an alpha source in the center with varied pressure of gas. And they were looking, if I remember rightly, at the gamma rays coming out of this object. This must have been in the Lise Meitner laboratory around 1930 or 1931, but I would have to look at the literature to know which.

Weisskopf: How do we get resonances?

Segrè: By the yield, you see, you find more gamma rays as you change the pressure. I remember this experiment but not when it was.

Telegdi: The ones that Maurice Goldhaber was talking about, I think, were just found from experimental observation. They had Boron 11 plus proton, Lithium 7 plus proton, and of course resonance was found very early.

Several conversations occurring simultaneously about early experiments make tape unintelligible here.

Weisskopf: The point that Maurice makes that Rutherford already thought in compound nucleus terms, is historically very interesting as it explains the fact that Rutherford understood Bohr so well, because, after all, understanding Bohr is not so easy. But it would be very interesting to find out somehow whether there wasn't an interaction between the two, whether Rutherford hasn't even induced some of the ideas.

Goldhaber: Earlier resonance.

Weiner: That can be checked by checking the correspondence immediately preceding this. It was interesting though, that although Rutherford gave an excellent report, it was Chadwick who commented that this brings us back to our earlier ideas. Rutherford didn't comment on it; it was Chadwick replying to Rutherford's description, saying, "Don't you remember," and bringing him back. So it's not clear that it is Rutherford's idea although the comments here sounded as if, in fact, Rutherford felt this way.

Goldhaber: He definitely talked of it in this way of sticking together for a while.

Weiner: How about elsewhere? Was there any other approach during the period that was similar, any other picture of the nucleus that was similar to Rutherford's picture.

Bethe: Harkins said that too without any justification, but you will find three or four papers by Harkins in which he essentially just writes down A plus B gives C gives D plus E. No justification why this should be, and as far as I read these papers, it only meant that this is a convenient way to add numbers.[52]

52. W. D. Harkins, in "Intermediate Nucleus and Atomic Disintegration in Steps," *Nat. Acad. Sci. Proc., 23,* 120-131 (February 1937), claims priority for the compound nucleus.

Uhlenbeck: But the chemists did that always.

Goldhaber: Did the chemists write reactions like this with an intermediate with no physical meaning to it?

Morrison: There is a physical meaning to it.

Bethe: Did they think that this compound held together longer than a vibrational period or something?

Weisskopf: Is it true in chemistry?

Bethe: It is true in many reactions.

Goldhaber: We are now doing molecular beam reactions like this in nuclear physics.

Wheeler: Speaking of premature things, we have spoken about Harkins. There was another premature thing, was there not, H. A. Wilson with the idea of rotational states of atomic nuclei, who had spectra with energy proportional to j (j+1).[53] He looked at the numbers. You know in some of the heavy nuclei you had enough levels.

Weisskopf: I remember always before Bohr-Mottelson in courses in nuclear physics that we calculated what it would be and found out it is much too small.

Segrè: That's right. Everybody made the rotational states of nuclei, absolutely everybody.

Goldhaber: You [Wheeler] and Teller.

Wheeler: Yes, but Wilson had really the observational evidence confronted with the calculations.

53. H. A. Wilson, "The Disintegration Energies of Radioactive Nuclei," *Phys. Rev., 44,* 858-859 (November 1933).

Goldhaber: Which time was that, roughly?

Wheeler: I think it must have been 1933 or thereabouts.

Weisskopf: But what kind of moment of inertia did he come out with?

Wheeler: He didn't worry about that. He just looked at the formula.

Morrison: And the states were probably from radioactive decay.

Wheeler: Wasn't H. A. Wilson originally from Cambridge?

Telegdi: There were two more guys who did these things.

Segrè: There was the paper of Teller and you [Wheeler] I think, on the alpha particle model of oxygen in which the period for overlapping was not one turn, but one-half turn, or whatever it was.[54]

Wheeler: Yes, right.

Weiner: Getting back to the lack of fashion of the shell model because of real difficulties in its early period, was there a certain question of style—a thing that we mentioned earlier—and taste, which determined whether one would look for the solution in a shell model or in a compound nucleus? For example, Professor Segrè made a statement regarding Fermi, that he would have liked from the beginning to see the shell model rather than the compound nucleus. What accounts for this?

Segrè: Fermi loved the shell model even before it was born practically and he never liked the compound nucleus.

Wheeler: You can see in fission he resisted the compound nucleus description of fission. He had nothing to offer in its place but he was struggling awfully hard.

54. J. Wheeler and E. Teller, "On the Rotation of the Atomic Nucleus," *Phys. Rev., 53,* 778-89 (May 1938).

Small: How widespread was this resistance?

Bethe: Wigner resisted it very much.

Shapere: Did he have any alternative?

Wheeler: No, there never was an alternative.

Weiner: Was it physics or esthetics?

Bethe: Wigner always said, I think, it was esthetics more than anything else. He complained that people looked at nuclei only in one way. He didn't say that the compound nucleus was wrong, but he complained that everybody worked on this and nobody worked on his symmetry ideas, that, after all, nuclei up to 40 or even more could be classified very nicely by isospin and ordinary spin and the extra quantum number and he had got, after all, very useful results from that. He predicted a few isotopes which had never been found, like sulphur 36. He said that ought to exist and it did. That all came from the symmetry analysis.

Wheeler: And Wigner also, I remember in early 1939, in objection to the compound nucleus picture would point out nuclei which differ with only two neutrons and stress that they had the same spin and the same magnetic moment, and this simply couldn't be in the compound nucleus picture. And he kept stressing this over and over.

Weisskopf: But I think one should not believe that the shell model didn't right away get acceptance. The shell model was like a bombshell. People right away saw that this is something.

Segrè: But when I said that Fermi liked it before it was born, it is because I remember at the end of Los Alamos when I was making all those big charts of isotopes, Fermi was always looking at them, and my hunch is that he had in the back of his mind something to find the magic numbers or something like that. He was too fast when Maria Mayer went to him to point out the spin-orbit coupling and everything. He must have thought a lot about it.

Bethe: Of course, we did talk shell model for the light nuclei. There were the calculations by Feenberg of everything between helium and oxygen which did predict some things.

Weiner: But you implied that there was ready acceptance of the shell model. Yet Professor Bethe has stated [in his interview] that as late as 1953 he proposed a certain test for the shell model hoping that it would come out wrong, and that, in fact, there was a reluctance to accept the shell model because it didn't seem to fit into things.

Bethe: I think I should amplify that. I think there was ready acceptance of the shell model as a guiding principle, giving magic numbers, giving angular momenta, and so on. What I didn't like in '53 was that it should be so rigidly true, that if you have 3 nucleons in a shell that you should necessarily have that third nucleon in a definite orbital. And it seemed to me that as in atoms there were many possibilities of coupling between the nucleons in a shell and that you could have different orbitals giving you the same total spin in the end. And I would have liked to see a nucleus having 3 or 5 nucleons in its outer shell [which] would be in a superposition of configurations, that is, not all in the same configuration. But it is.

Goldhaber: What was your specific proposal at the time?

Bethe: It was the question of a stripping reaction and there were two nuclei—I forget which, but it was around chromium—which could go from one to the other by stripping, adding a small amount of angular momentum. And I think the question was: Did you add an $f^{7/2}$ particle when, according to the angular momenta, you could just as well have added a $p^{3/2}$? It turned out it was $f^{7/2}$.

Segrè: There was, at the beginning, though, the idea that the shell model would work only for the ground state and perhaps the first excited, but nothing beyond that.

Weisskopf: In some ways, this is true. It's getting worse and worse if you go on.

Telegdi: Also, I would say the shell model at the beginning was full of semi-empirical auxiliary rules, just like Professor Bethe referred to—four or five extra nucleons. It was really full of magic—what you had to do. There were not too many who said more results and less rules, but still I imagine there were people perturbed by the number of auxiliary rules.

But what I remember best from those days is Wigner's opposition. He came to Chicago when Maria Mayer was there and he was as anti-shell model as you can be, but partly for completely different reasons. He didn't like the idea of an *ad hoc* spin-orbit force. He had all these things work by tensor forces. He didn't like the connection between this empirical spin-orbit force and what one knew from nuclear forces obtained from nucleons in scattering experiments.

Bethe: I agree with him on that point. It was only much later that you saw a spin-orbit force in scattering.

Morrison: That in polarization.

Weiner: Am I correct in concluding that there was no real tension in the co-existence of several models except on the part of people who were firmly antagonists of one or the other?

Weisskopf: There was a time when people groped with the problem of the co-existence of the shell model and certain well-established compound nucleus results, whether this was a contradiction or not. It took some development before one could see that it is in fact not a contradiction.

Small: Was this more of a psychological problem than a physical one? What I mean is: Is there more of a willingness today to live with a variety of models?

In Unison: No, no.

Morrison: Not in this non-fundamental sense.

Weisskopf: It is really true that the two effects can both exist at the same time.

Wheeler: But isn't there also here, just to develop that comment of yours that it's partly a matter of temperament, that there was a considerable group of our colleagues who felt very unhappy with anything like a compound nucleus model. It wasn't the style in which they were accustomed to do physics. You didn't write down a Hamiltonian, and then have a perturbation, and do things in the usual systematic way. It was something that was highly physically intuitive. And was it not true that all during this period before there was a shell model, people with this temperament that I'm describing who felt uncomfortable with the compound nucleus model were really resisting it completely? I don't know any model in nuclear or in any part of physics which went longer unaccepted by a larger fraction of the community than the compound nucleus picture. And those were people who were waiting for the shell model to come along.

Segrè: Does anybody know Bohr's reaction to the shell model?

Weiner: Here is where we really need the documents.[55] The letters we have are up to 1945.

Wheeler: Yes, I feel so stupid, because in 1949, I was making periodic visits to Copenhagen and we were trying to develop this unified picture of the collective model. And I can remember coming back on the train from Copenhagen to Paris and then suddenly getting this idea of the particles going round and pushing on the nuclear wall. This was, of course, something that had already been done independently and really was on its way to being published at that moment by Rainwater.[56] But here we were working on this paper which was to unify the independent particle picture and the compound nucleus picture. He felt very strongly the need to put these two together. It

55. Bohr's reaction to the shell model is discussed in H. D. Jensen, "The History of the Theory of the Atomic Nucleus," *Science, 147,* 1419-1422 (1965).
56. J. Rainwater, "Nuclear Energy Level Argument for a Spheroid Nuclear Model," *Phys. Rev., 79,* 432 (August 1950).

was not something that you could laugh off and stay away from. You had to plunge in 100%.

Weisskopf: I remember that discussions with Bohr brought me and F. L. Friedman to this idea that when a particle comes in and it goes out right away, then its behavior is determined by the shell model—a simple particle potential is correct for that description—but when it stays in the nucleus for a long time, then it is the compound nucleus picture that will be right. The purpose of our paper was to get these two things understood and to show that they are not contradictory but actually the right description. A discussion with Bohr helped us, in particular to get the time element in. If you have short time processes, you have large energy intervals to look at and you average over energies. When you have long times, you get the fine structure. This was clarified by discussion with Bohr. In other words, I support your statement that Bohr has helped—at least you and me—to get some bridge between, to understand both phenomena.

Bethe: What was your paper with Friedman in?

Weisskopf: That was in the Bohr Festschrift in 1955 when Bohr was seventy.[57]

Wheeler: This paper with Hill and myself in 1952,[58] that was really Bohr's paper in the sense that the ideas had come out of discussions with him, but the H-bomb business had separated me from him, and somebody had to write it up and it ended up with the two of us writing it up. But really this was his reaction to the necessity of having both the shell model and the compound nucleus picture.

Weiner: Apparently there was on the part of some people and some groups a

57. F. L. Friedman and V. Weisskopf, "The Compound Nucleus," *Niels Bohr and the Development of Physics:* Essays dedicated to Niels Bohr on the occasion of his seventieth birthday, edited by W. Pauli, with the assistance of L. Rosenfeld and V. Weisskopf (New York: Pergamon Press, 1955).
58. D. L. Hill and J.A. Wheeler, "Nuclear Constitution and the Interpretation of Fission Phenomena," *Phys. Rev., 89,* 1102-45 (March 1953).

conscious program over a long period of time to reconcile the models. We're talking about theorists. How about experimentalists? Were they caught up in this in any way? We heard about the proposal of a theorist to do an experiment in 1953, which you [Bethe] hoped would come out against the shell model. Were there any other instances of this, where groups of experimentalists were involved?

Wheeler: Rainwater on the quadrupole moments of nuclei.

Weisskopf: The experimentalists found processes called direct reactions at the time.

Segrè: When were the Butler strippings?[59] In what year were they?

Telegdi: In 1950.

Bethe: Yes, that's about right.

Telegdi: He was a student of Peierls.

Weisskopf: Who introduced the term "direct reaction?" Do you remember?

Weiner: Was this the same quality of debate and discussion and desire among experimentalists though? You have given me instances of people being involved. But would you say this was all of physics?

Bethe: Did Barschall do this consciously in order to . . .

Weisskopf: Hans Barschall wanted to destroy the idea of the compound nucleus when he made these angular dependences.

Bethe: These were experiments with medium-energy neutrons, a few hundred

59. S. T. Butler, "On Angular Distributions from (d,p) and (d,n) Nuclear Reactions," *Phys. Rev., 80,* 1095L (December 1950).

kilovolts, finding variations with atomic weight which were very regular. Beautiful experiments.

Weisskopf: Beautiful experiments over a long time made by Barschall and collaborators, and which later on gave us the opportunity to test this intermediate model.

Weiner: But his motivation, you say, was to disprove the compound nucleus?

Weisskopf: I believe so, but he is still there to be asked.[60]

Telegdi: There was a tremendous amount of upsurge of renewed interest, as Maurice [Goldhaber] will tell you, in isomers. That was the great thing. Look for isomers, and the shell model will tell you how many there would be, what their spins would be.

Goldhaber: There were always some isomers which shouldn't be there in the shell models, and then on careful investigation they disappeared.

Bethe: I think there was a lot of activity directly connected with proving the shell model or disproving it, not so much with reconciling the two models.

Telegdi: This had a tremendous effect on beta decay because people started working out $F\tau$ values and taking them much more seriously than they ever had before. That was the tremendous influence of the shell model at that time.

Shapere: The people who were interested in reconciling the two, did they make any explicit connection with complementarity?

Weisskopf: Well, yes, in a way, but it has not much to do with complementarity.

60. See H. H. Barschall's discussion of his early work in "Three Decades of Fast Neutron Experiments," *Physics Today, 22,* 54-59 (August 1969).

Bethe: The thing you explained before is about that connection.

Weiner: What interests me about this is that the upsurge in interest and sense of renewed excitement comes at precisely the time when particle physics is beginning to get very exciting and to form as a separate field. And so, if the historian wanted to make some hypothesis and say, well, nuclear physics was getting dull and therefore people wanted to do particle physics, it wouldn't be true.

Telegdi: No, no, it wasn't getting dull, and also in many parts of the world—I can speak here for a first rate laboratory at that time in Zurich which, of course, had no high energy, neither equipment nor tradition. Nuclear physics had become, as you know, something people knew as very important to work on, but much of the work, say in nuclear spectroscopy, that went on was in no way related to anything that you could test. And then all of a sudden when the shell model came out, all these people had a purpose. It became much more specific. They had a systematics in beta decay and gamma ray energies and so forth. Since the shell model came out, these spectroscopic measurements had acquired much more significant meaning.

Weiner: What about people who had already moved over into particle physics?

Weisskopf: Maurice [Goldhaber] just made a profound statement to me. Fermi discovered the famous 1235 MeV resonance in 1951, just about at the time of the shell model. Had he not been interested in mesons, he probably would have worked out the shell model himself.

Goldhaber: Or at least worked on it.

Segrè: This I know quite well. Fermi, at the end of Los Alamos, said, "Nuclear physics fits into place like spectroscopy twenty years ago, I'm going to go to particle physics. I'm going to start studying." This is absolutely sure.

Weiner: Like Corbino in 1929.

Segrè: He repeated the same operation that he had done in 1929. Of course, he was interested in anything. The proof is a little book[61] where you find all the calculations that he had prepared. He had prepared absolutely everything he could do with phase space and without knowing anything to interpret the high-energy data as soon as it would come out. And he was waiting for the machines.

Small: I'd like to raise a point just mentioned here by Dr. Segrè, and that is: In a recent interview, Professor Weisskopf has stated that atomic physics was more or less done with, or finished, I think he said. We have a similar statement from Professor Bethe that the work on the atom was done around 1930. I don't know exactly what you mean by this, but is there a similarity between that kind of a finishing off of an area and the state in which nuclear physics found itself after the War?

Bethe: No.

Small: In what sense then was atomic physics finished?

Bethe: We knew the forces, we knew quantum mechanics, and I think everybody was convinced by 1930 that if you had enough mathematical tools you could calculate the levels of any atom with any accuracy.

Wheeler: Dirac's famous statement that everything was known.

Bethe: And this state in nuclear physics has not been reached today.

Small: So I assume this is not what Fermi meant when he decided to go into particle physics.

Morrison: In a tactical sense, it is what he did mean.

Weisskopf: There is another analogy, I think. It is the analogy that Rutherford left atomic physics in 1917 and went into nuclear physics.

61. Enrico Fermi, *Elementary Particles* (New Haven: Yale University Press, 1951).

Morrison: 1917? 1897.

Weisskopf: Rutherford has discovered the structure of the atom.

Segrè: By accident. He was making nuclear physics.

Wheeler: I thought you were going to speak of the analogy when Rutherford was needling Bohr to get his paper on atomic structure finished, and Bohr said, "Yes, but I've only explained hydrogen. I have to explain all the atoms first." And Rutherford said to him, "You explain hydrogen and you explain helium, and people will believe that you could explain all the rest." Of course, he never even explained helium.

But here is an analogy. Atomic structure was explained, in Fermi's view, nuclear structure was not explained, but you could think of it as being explained, so you could go on in good conscience to something else.

Weisskopf: I think that Hans's statement was a little too exaggerated. I agree with you that the situation after the War was different from 1930 in atomic physics because we don't have the elementary force. But there was some kind of armistice there in nuclear physics in the sense that you this morning described as sort of an intermediate position. It was clear at that time that if one measures the nuclear force reasonably well, then one will be able to calculate most of the nuclear properties reasonably well. And I think this is probably what Fermi meant when he said, "I'm now going into particle physics," because he probably felt that the real solution of the nuclear force is not going to be there, it is going to be in particle physics.

Telegdi: I think also Fermi liked in general to create his own experimental facts, and there were too few facts about mesons but lots of talk.

Segrè: I remember also, much to my surprise, to have heard from Bohr at the end of Los Alamos, if I understood him right, the idea that physics was about to finish.

Goldhaber: He didn't mean physicists?

Weisskopf: How did he argue that?

Segrè: He said to me that when one knew neutrons and protons in nuclei, there was not much left to find.

Weisskopf: Did he change his mind when he found the resonance?

Segrè: I mean Bohr.

Weisskopf: I'm sorry. I thought it was Fermi.

Segrè: I am sorry. I have confused now the two things. Bohr, I remember if I understood him right, saying that he thought that physics in general was going to finish. And I was very surprised to hear Bohr say anything like this. But he said, "Well, one knows neutrons, protons, electrons, nuclei and everything else, and it is going to finish." He didn't mention particle physics at all. This was the end of physics, he said.

Bethe: Wasn't he aware of mesons?

Segrè: I am very reluctant to vouch for what I understood from Bohr because it is very difficult for me to know what he said. But this is what I understood at the time.

Morrison: Did Copenhagen know anything about mesons from 1946 until 1955?

Weisskopf: It is true that Copenhagen was a little further away from meson physics.

Telegdi: Powell said to me he constantly kept well informed about it.

Goldhaber: This was a little later. He said that at the end of Los Alamos.

Segrè: It was at the end of Los Alamos. He may have changed his mind afterwards. I'm speaking now of 1945. I left Los Alamos the very end of '45.

Weisskopf: He certainly changed his mind. He was instrumental in founding CERN.

In Unison: That is much later.

Weiner: This raises the question about the effect of the war period on the subsequent development of theory. It has come up several times—about people getting ideas then, or deferring ideas that they had beforehand—and I think that maybe we can consider this as the time for it. In the context of the things that we've been talking about—models, forces, the beginning of particle physics, the hopes for understanding the nucleus—what was the effect of the wartime experiences at Los Alamos? I'm not talking about the sociological experiences, the changes in funding, or the group experiences, which are a story in themselves—but the theoretical discussions there in the physics seminars (not in the actual technical project) which were held during the war years. Especially during the retooling period when people were beginning to have refresher courses and thinking about the theoretical problems in so-called pure physics that they were once again going to tackle. Can you identify any significant developments during the period? By the way, it needn't be only Los Alamos but I am using that as an example of the wartime experience.

Morrison: Hans went to G.E. in 1946 and there gave this very successful review—a post-war re-do, one might say.

Bethe: But there was nothing new, nothing at all. That was just to bring a new generation to the point where we had been in 1940.

Morrison: The deuteron problem was perhaps a little more carefully done because of the many things that went into it.

Weisskopf: And the statistical theory was more carefully done. The point is that the fission process, as Johnny [Wheeler] and Bohr had shown, is well described by the compound nucleus picture, and also the inelastic scattering phenomena. All these things that were very important for the bomb development are all reasonably well described by statistical theory. Therefore, what happened in Los Alamos was a finishing up and also a going into depth of statistical theory. And that may be the result.

Wheeler: Were you present at Los Alamos at the seminar where Feynman talked of the work on half-advanced half-retarded potentials and "sum over" histories and so on, and Bohr was violently opposed to this idea at the seminar.

Weisskopf: I'm talking now only of nuclear structure physics. You bring up a new thing. The whole new developments in electrodynamics may have started there, but I thought that was not Charles's question.

Weiner: I'd like to get into that too later.

Segrè: At Los Alamos there was a tremendous accumulation of materials which didn't exist before and which made the shell model. The shell model could not have found experimental support before Los Alamos.

Weisskopf: But it did not grow out of it.

Segrè: Everybody looked at fission chains.

Weisskopf: I believe the most typical example for the triumph of the statistical theory is illustrated by the following incident. The first measurements of the average neutron cross section in uranium was very low at low energy; it was one-third of the statistical cross section. I think it was I who insisted that the statistical theory must be right, and the experimenters laughed and said it is a factor three wrong. Later it turned out that the measurements at only slightly higher energy gave a higher value and showed that the low value at low energy was a fluctuation. This characterizes the predictive power of the statistical theory. It was the great heroic time of statistical theory.

Wheeler: In the statistical theory we also remember how well the predictions came out for the number of gamma rays and how they fell off with the time after fission—"delayed energy release."

Weiner: What about the other theoretical developments? I'm glad you did separate them from nuclear structure, but it seems to me if we're going to understand the subsequent postwar switch to particle physics on the part of many people we have to take it through a quantum electrodynamics stage.

Segrè: No, the phase stability, don't forget that, so that you could bring it in the laboratory.

Bethe: I think this was the important thing. Fast detection evolved as a Los Alamos product.

Morrison: I remember so plainly before the war watching people building amplifiers and if it was up to 50 kilocycles they were very pleased.

Segrè: This was a spectacular thing. I remember having an amplifier and an ionization chamber and trying to measure alpha particles with tremendous efforts. The ionization chamber was full of air, and then we put in nitrogen. I read a paper by Ortmann and Stetter in which you collected electrons instead of ions and reduced the collecting time by a factor of a thousand. This was something unbelievable, without changing the apparatus or anything. Just putting another gas in.

Morrison: As long as you had a wide band amplifier you could take advantage of it.

Segrè: Nothing was microphonic any more. This was the introduction of the electronic age.

Bethe: So it was really advance in instrumentation, not in theory.

Morrison: I would say it was instrumentation in the large—instrumentation plus group work plus experts in various fields. The whole style of the war work brought along in its train lots of data, good instrumentation, new accelerators, and the idea that you could work all these things together.

Wheeler: The spontaneous fission and the fact that it varied so much between even and odd—that was very important in pushing more modern nuclear theory to the foreground because that didn't fit with the elementary compound nucleus.

Weiner: Something interesting that came up that's a little bit off the track, but twice today Bohr's criticism of Feynman was mentioned, of the particular style of presentation perhaps.

Telegdi: Style of thinking.

Weiner: And I can think of a third instance at the Pocono Manor meeting [1948], one of the so-called Shelter Island Conferences, where again there was this criticism. Does this represent this question of personal individual differences in style, or is it a question of a different way of doing physics that was just out of Bohr's boundary conditions?

Wheeler: I think the word "criticism" only meant that here he felt a worry about how was one going to bring this way of thinking about things, whether it was Feynman's way or somebody else's, into the ball park of firmly established principles and as something that could be discussed in terms of observation and principles of measurement. Because, just to give a formalistic thing without making a deeper analysis of it, was foreign to his approach. In other words, I don't think it was so much criticism as indication of an incomplete understanding.

Uhlenbeck: Feynman took it very hard. I remember at Pocono that Bohr said to Feynman, "You can't talk this way, you can't talk this way," and Feynman was simply shut up. He had worked very hard and I think the main point was that his way of talking was, of course, unconnected with what the previous people did. And it was, in a sense, too original. Therefore, he didn't object against Wigner because that was immediately in the old tradition.

Bethe: And I think what John said did come into this, namely, Feynman used the world line going backwards and forwards, and Bohr, of course, would not accept this without thinking for a year of what it would mean.

Telegdi: As I understood it, he didn't like the world line even going forward because then he found that you couldn't tell the position and the momentum of particles simultaneously—you should know that. That's a long way back though.

Uhlenbeck: But still Feynman took it so seriously that he connected it with the knowledge, and at the last conference which was at Oldstone-on-the Hudson [1949], he dominated the proceedings. Bohr wasn't there though, but at least everybody was convinced.

Bethe: That made a difference that Bohr wasn't there.

Uhlenbeck: No, I don't really think so because he had worked so hard to show that you could look upon it as just another mathematical matter. I think everybody was completely convinced at that time.

Weiner: This raises a question about the series of three conferences, the first in 1947, and then the two in the following years. The subject of the first was something like Fundamental Problems of Quantum Mechanics, and that was where the Lamb shift was introduced and discussed. I would like to take these three conferences, either together or one by one, and relate them to our subject, which so far has been "The Role of Theory in Nuclear Structure." Was there anything relevant at these conferences?

Bethe: Yes, the shell model.

Uhlenbeck: But not very much. I believe the main emphasis was quantum electrodynamics.

Weisskopf: And particle physics. Those were the two main subjects. Also pions.

Bethe: Oh yes, that was very important already at the first conference.

Uhlenbeck: The two mesons.

Weiner: Did these conferences represent any sort of growth in terms of theory? It seems to me rather interesting that Oppenheimer described them as the last intimate conferences of physicists that he could remember, small groups. They, as a matter of fact, only lasted for three years, and then the Rochester conferences began which became just the opposite, very large ones.

Bethe: It wasn't so large to begin with.

Uhlenbeck: It was still larger than the others.

Segrè: But also, the other ones were purely theory and at Rochester there were also experimentalists.

Bethe: We had some experimentalists. Rossi, wasn't it?

Weiner: People were invited if a particular subject was hot. But what I'm trying to do is to see if there were any links in the problems and the concerns expressed on the agendas and in the discussions of those conferences, with the transition that later took place with the blossoming of particle physics as a very rapidly growing, separately defined field.

Weisskopf: Is it not so that the Shelter Island conferences have shown that the leading people were getting away from nuclear structure? Perhaps I am expressing it too strongly, but Oppenheimer, Fermi, and to some extent, Hans . . .

Bethe: It was particle physics and electrodynamics.

Weisskopf: I would like to express it this way. As a young man, one would get the impression, if one had listened to those conferences, that the thing to do nowadays is electrodynamics and particle physics, and no longer nuclear structure.

Weiner: That is the impression the historian gets from outside in looking at it, and I just want to know how correct that impression is.

Uhlenbeck: With electrodynamics, I almost call it the "renormalization generation." It was quite clear that around three or four years it looked as if everything would go suddenly again as in 1925, and since physicists are so psychologically brought up with the "break-through" idea, here it suddenly seemed as if it would go again. As a result there were quite a lot of very bright young people—students of Schwinger, students of Oppenheimer, and also at the Institute [for Advanced Study]—who were completely absorbed by that. You can name them—Pais, Watson, Karplus . . . . If I would think, certainly I could name ten of them who at that time made their career. That was the thing to do and they made contributions at that time, just as I was brought up in the quantum mechanics generation. When something like that happens, then suddenly it attracts a lot of the brightest people. Now it petered out. In three or four years it was over.

Weisskopf: It petered out much faster than quantum mechanics.

Uhlenbeck: Yes, and it was in that sense a false alarm. But still, in regard to education, it had the same effect, I think, and as a result, after 1950, it was clear for a young theorist that he must, of course, do this, particle physics or high energy physics, and not more nuclear physics, because that was considered doing chemistry or something like that.

Bethe: Yes, I think nuclear physics was quite dormant between 1945 and the shell model.

Weiner: And that's precisely when these conferences took place.

Wheeler: The compound nucleus picture perhaps made it seem more like chemistry than it would have done had the shell model been in existence.

Weisskopf: On the other hand, dormant is perhaps not the right word. This is the time when Blatt and I wrote our book.

Segrè: Well, it snored very loudly.

Weisskopf: Exactly. You also edited a series of books. There was a large number of students, in fact, the classes were enormous in nuclear physics and in nuclear structure. Dormant is the wrong expression. There was some stagnation, but I wouldn't call it dormant.

Morrison: This is the fulfilment of the missing years of the war somehow. They all came out together at this time and a great many experimentalists did nuclear physics.

Weiner: Now I know the meaning of Dr. Uhlenbeck's term "renormalization generation" in another sense.

Morrison: I would think that there is really a logical difference. I simply can't help thinking—perhaps this is unkind—but I can't help thinking that nuclear structure *is* more like chemistry, just intrinsically so. No matter what you say about it or how ingenious you are with shell models, it is not any longer quite the same thing as even the deuteron problem.

Wheeler: One wonders if today elementary particles are beginning to be looked at as more like chemistry than electrodynamics.

Morrison: I believe they are.

Weisskopf: Are you making the same point I tried to make that in some ways the fundamental question has moved on and one hopes that the nuclear structure questions will be solved by answering the fundamental question, in principle.

Morrison: In principle solved, yes, and that another set of people, perhaps closer in some ways to the details of the subject, have to translate the "in principle" into practice, just as certainly chemistry has not, in fact, profited very much by Dirac's book.

Weisskopf: But by quantum mechanics, it has.

Uhlenbeck: There is still much to do.

Morrison: There is plenty to do, yes.

Uhlenbeck: Chemistry is in principle explained, but that principle is in heavy quotation marks.

Morrison: It just happens if your interest is that way, and I think I share it. But to be honest, those people have many ingenious calculations and conceptual approximations very like the shell model and so on. It's on the same level. It's this intermediate ground which Hans spoke about.

Bethe: I still say that the status of nuclear physics in the four years directly after the War was somewhat unusual. Certainly nuclear physicists were waiting for the machines to be finished so that we would learn more about scattering and about nuclear forces, and that we would learn more about particles which might give us the basis for the nuclear forces. At the same time, nuclear physics had shown itself to be of very practical importance so that lots of people wanted to be trained in the field.

Uhlenbeck: Giving up engineering.

Weisskopf: Where the dormant element is inappropriate.

Bethe: I concede.

Weiner: If we wanted to examine that period—I think my feelings are confirmed that it is a crucial transition period—one of the ways that we thought about is to take a look at the Shelter Island Conferences themselves as symptoms of some kind of a change, and that is possible because Oppenheimer's personal papers preserved at the Library of Congress contain the records of the Shelter Island Conferences, the original letters on it, including a letter from you [Weisskopf] to him regarding Schwinger's work, and some very interesting things with discussions about the agenda. Kramers, I think, was involved too. And a letter from Dirac saying it was the best conference he ever attended. But there are problems and there are ways of getting at them. I was wondering if we're taking a look at that period, whether this would be one good way to approach it.

Weisskopf: If you look at nuclear structure in this period, my memory is gone, but I still believe we'd be doing it an injustice to say that nothing has happened until the shell model—the giant resonance, all the understanding of gamma ray transitions. Hans, I think it is wrong to say that nothing has happened in nuclear structure.

Morrison: Were any angular correlations coming in?

Segrè: Angular correlations were before Los Alamos because I remember Deutsch had already made a reputation because of the angular correlation when he came to Los Alamos.

Weisskopf: There must be much more in that time. From 1945 to 1951 certainly nuclear structure had a tremendous development.

Segrè: Because the empirical material had grown so much the systematic studies which were impossible before . . . For instance, take the alpha radioactivity, the extension of Geiger-Nuttall's Law, things of this type—once you had three times as many substances to look at.

Weisskopf: It was also conceptual. I mentioned the giant resonance. There must be more.

Wheeler: The deformed nucleus was the single most important step in that period. That was 1948 or 1949.

Segrè: The electric quadrupole moment of the nuclei is before the war.

Bethe: Yes, and there were a lot more in this period.

Weiner: I think that defines a project for us—for a specific look at the literature and at the subjects of the doctoral dissertations—what people were doing and what people were saying. I think we can learn a little bit about that before you come back next time.

Bethe: There must be a lot in your book. [Blatt & Weisskopf, *Theoretical Nuclear Physics*]

Weisskopf: The book contains the whole thing up to the prewar development.

Bethe: And you have references.

Weiner: I was thinking of looking for the symptoms of what was going on—these little indices, these social indicators of relative sleep, loudness of snore, or relevant position in regard to other fields. I find it an interesting period.

Weisskopf: The development of weak interactions of the different interactions after KU. [Konopinski and Uhlenbeck]

Uhlenbeck: Forbidden transitions.

Segrè: That was a big thing after the war, the big elaboration of V - A.

Morrison: The behavior of free neutrons was done then, in 1950.

Uhlenbeck: There was experimental proof.

Bethe: And, of course, on the systematics of beta decay, there were just so many more substances available—there it applies even more what you said about alpha decay—and in such bigger quantity, and so much better detecting equipment that you could study all this. The neutrino . . .

Segrè: Oh, Allen. Allen was after the war when he left Los Alamos and went to Illinois.

Bethe: And that was very soon after the war, I believe.

Morrison: It had been tried before the war, you know.

Bethe: Not very successfully. The Allen experiment was the first really successful experiment. Reines was much later, '56.

Weiner: I would like to suggest a procedural point now. I think we have covered a lot of ground on this type of discussion. There is a final point on the agenda relating to evaluation of this conference itself as a technique, about future steps in our project, and about other kinds of conferences, interviews, documents, studies, substudies—not necessarily things that the few of us who are putting some time into it can do, but things that ideally should be done.

Before we get into that, I wonder if there are any points that we feel are important to resurrect, points that we have missed, major points. Obviously, there are many. But is there something that you would like to raise now before we go on to this final change of pace?

Bethe: One point I think you should investigate is the history of the stripping reaction—the deuteron leaving either its neutron or its proton in the nucleus. This is a postwar development. The Butler theory I think we said was 1950, but experiments had gone before, so this must have been a live subject experimentally.

Morrison: Of course there was the work of Oppenheimer-Phillips before the war.[62]

Bethe: Yes, but the use of the angular distribution was new, and I think Butler's paper must have been preceded by the measurement of peculiar angular distributions and I just don't remember that.

Segrè: Was something done at Liverpool? Because when I was still at Los Alamos somebody showed an angular distribution made with photographic emulsions at Liverpool—I remember that. I think Chadwick showed it but I don't know precisely.

Uhlenbeck: Parkinson did it at Ann Arbor quite a long time, for many years. It was also in the fifties.

62. R. Oppenheimer and M. Phillips, "Note on the Transmutation Function for Deuterons," *Phys. Rev., 48,* 500 (September 1935).

Segrè: Wasn't Chew's doctoral thesis on stripping?[63] Yes, the idea of the spectator nucleus was in Chew's thesis, I know that.

Bethe: It could be. Certainly this became tremendously important for the confirmation of the shell model because you could tack on individual neutrons and protons with angular momenta, and the origins of this would be very interesting.

Weiner: Actually, we would have to have a different group to talk about it and to give us information on this immediate postwar period.

Segrè: If you just look at Butler's book—there's a little book by Butler in which these things are recorded.

Bethe: Another interesting problem which we have touched many times is deviations from the compound nucleus. You mentioned Barschall's experiments but there were also Cohen's experiments.

Weisskopf: Yes, actually I think Cohen is the one who introduced the direct reaction.

Bethe: Yes, and he went to every conceivable meeting and said the compound nucleus is wrong. Look at my experiments.

Wheeler: Another point is the existence of two different cloudy crystal ball pictures, one as applied to low energy nuclear processes, and the other of Fernbach and Serber which applies to high energy processes.[64] For a time these co-existed, as far as I could see, without them being really ever brought together as one single model that was being applied at different energy regimes. Serber was commenting on how it was really Karl Darrow who introduced the word "cloudy crystal ball." You were presiding, I believe, at a

63. G. Chew, "Elastic Scattering of High Energy Nucleons by Deuterons," 1948 Thesis.
64. S. Fernbach, R. Serber and T.B. Taylor, "The Scattering of High Energy Neutrons by Nuclei," *Phys. Rev., 75,* 1352 (May 1949).

meeting where he gave his paper, and then Karl Darrow, at that point, used this word.

Weisskopf: Well, the Fernbach-Serber cloudy crystal ball happened in that dormant period in 1949.

Bethe: That was in response, wasn't it, to . . . .

Segrè: Our experiments.

Bethe: The high-energy experiments.

Weisskopf: Then the concept of strength function.

Bethe: Isn't that later? That followed your paper, I think.

Weisskopf: Maybe that's right.

Weiner: Any other points, not only on our last discussion, but on the entire range of things we've covered, or things of that type, that you feel we can explore successfully now?

Wheeler: Let me just add what a sense of humility it induces to see how ignorant one is of the literature and how little one remembers. Being here makes me, at any rate, feel that I ought to study for a month ahead of time.

Segrè: We have said quite a little about beta decay and there was a real big enterprise to pin down the interaction in beta decay which lasted many many years.

Morrison: It was wildly unsuccessful. It alternated back and forth. It was after the war. I remember very well at Cornell. It certainly went on from 1947 or 1948. There were six experiments which turned out to be wrong.

Segrè: V and A. And V - A. And S - T. And so on.

Weisskopf: This was actually in the dormant time, because I remember that was in the book.

Weiner: Dormant time is becoming less dormant all the time.

Weisskopf: That is my point.

Caws: I would like to raise a second order question about the relationship of the people here to the whole movement of ideas as it was taking place. There is a certain relation of exteriority now conferred by time. We're talking about something that is not now happening. It is the object of an inquiry on your part and on all our parts. But I wonder about people who participated in these movements, to what extent they themselves had the sense of being spectators of the movement and to what extent they really had the sense of being participants in it. That is to say, to what extent does the historical development really go through the conscious reflection of the people participating in it and to what extent is it something which has an existence independent of their conscious participation in it? Obviously, the memory of these events is somewhat selective—some things stand out, some are no doubt of central importance, some may be of more anecdotal value and not really central to the story—but I wonder if the story could be written without its participants, just from the spectator's point of view?

Weisskopf: Who is the spectator?

Caws: I'm asking whether, for example, you yourself when working, when immersed in the problem—not now as you are at a distance of time looking at it as it were from the outside—but when actually immersed in the problem, to what extent did you have the feeling of watching something happening and to what extent did you have the feeling of actually making something happen?

Weiner: In other words, the simultaneity of being a participant and a spectator.

Klein: I think they are complementary states.

Uhlenbeck: Maybe I can speak for myself alone, but in those times you just did your damnedest. With regard to me, one of the great questions was that you had students, and you had to find theses for them. So that was an enormous problem all the time. Of course if there was something opening up, then you jumped on it, because then there were chances that you or your students could do something. In that sense, you participated, but I can't say that you were a spectator. You were completely immersed in it.

Caws: Sure, but what you say for example as to something opening up and you jumped on it, there's a kind of exteriority there—watching to see what happens and then taking advantage of certain things that happen.

Uhlenbeck: You looked at the papers, of course, or you went to a meeting and you heard them discuss something there. So I don't see how you can make a distinction between the two.

Caws: Maybe you can't . I was just raising the question.

Weiner: I know what he means. I mentioned that in some interviews I had the feeling of being a participant and a spectator. Historians do tend to think this way. But I've had physicists describe to me pieces of work that they have done—fundamental things that have been recognized by their colleagues as such. And I have asked this question: at the time of doing it, did you have a feeling of the impact of this work? And in one case in particular, I remember, a person said, "Yes, I knew exactly. I could visualize the abstract of the article. And I said to myself, 'The boys will like this.' " This is in the sense of knowing the significance of the work as you're doing it. When you come up with the idea, you know that in fact it is important, it will make a change, and it will be appreciated. And in that sense, I think, it comes closest to this participant-spectator duality or simultaneity.

Caws: I didn't really mean understanding the import of the work for the field but simply the relationship of the investigator to the work. You know, he's doing a bit of it. He sees the developments taking place elsewhere indirectly. He gets papers from other people, and so on. And he is making, perhaps knowing it or not, a fundamental contribution to the field. It isn't so much, as I say, the apprehension of oneself as indispensable to the development of the field but just the conceptualization of the field as a whole that somebody who is actually working in it has. Clearly 99% of the work is being done by other people. There is an exteriority there.

Weiner: I think that this consciousness of the whole field, where it is going, and what part you're playing in it, comes at a later day. With responsibility for great resources, funds, students, directing essentially 30 people's lives, pointing men in future career directions, spending other people's money—then you need to be conscious of it, and it becomes built-in at that stage. But you didn't really ask me, did you? Are there other general points of the type we have discussed already before we move on to our suggestions for next steps?

Buck: I'm interested in the problem of group work in physics, and I've especially looked at the school at Rome under Fermi. Professor Segrè was there and I have read a lot of his materials. I was wondering if there are any other cohesive groups in the thirties that could be compared to Rome, either in experimental work or in theory?

Bethe: Berkeley . . .

Weisskopf: Niels Bohr . . .

Segrè: The Cavendish . . .

Buck: Were the projects—were there three, four, five-author papers, for instance, coming out of those places?

Weisskopf: Hans Bethe's Cornell group.

Bethe: Berkeley certainly had lots of papers with three, four, five authors, the Cavendish usually two or three. In Copenhagen, I think it was not customary to have many authors.

Weisskopf: I'm not sure what you mean. Do you mean papers that actually are signed by many people, or do you think of work where people stimulate each other by very close contact? It seems to me that who signs the paper is less important.

Buck: Yes, that's quite true. In the case of Rome in the experimental work there was a sort of division of labor that Professor Segrè has described. Perhaps signing the papers was more relevant there. But in theoretical work, the colloquia and lunches and afternoon walks and so forth are a different thing. So that in Copenhagen there was certainly strong interaction without signatures appearing on papers.

Segrè: Heisenberg in Leipzig must have been at that time.

Weisskopf: No, to a lesser extent. One can never make a direct statement, but I believe to a much lesser extent. I think not as much as in Copenhagen or in those centers like Berkeley or Cornell. There was less give and take.

Segrè: In Rome there was one thing—a very small difference of age. Fermi, the oldest, was born in 1901. Amaldi, the youngest, was born in 1908 or 1909.

Weiner: And it was primarily a research group, not a teaching group.

Segrè: It was everything because all the physics had to be taught within the group from each other and essentially Fermi . . .

Weiner: No, I mean, teaching responsibilities to large groups of graduate and undergraduate students. It is quite different whether you are talking of groups or whether you are talking about people who happened to be clustered around an institution. For example, the thing that I learned from the Chadwick interview, and probably it should have been obvious, was that basically the Cavendish was really a teaching institution. If you really look at it, in fact, it was the greatest teaching institution in nuclear physics in the world for a long time because you had people coming for their doctoral work and for postdoctoral work. They had 30, 40, 50 students, and they would go to Chadwick's attic course, a basic orientation course that he'd run for them to teach them not to break too many things and to teach them the rules (not the rules about going home at the proper time) so that they could handle the sources and work with the instruments. It really was a teaching laboratory and that was its primary purpose. And Rutherford saw it that way. The research was bonus.

Segrè: That was the Cavendish, but Berkeley was a pure laboratory only for developing the cyclotron, and if you were a good boy you were permitted to do some research. But this was just a privilege given.

Morrison: That was on the experimental side.

Weiner: Yes, that was the experimental Berkeley.

Wheeler: DuBridge couldn't have looked at it that way.

Bethe: Oppenheimer's group at Berkeley was a teaching institution, and, at the same time, had a number of postdoctoral fellows . . .

Morrison: I think it was characterized by a very wide set of interests—a little like Landau's, in a way, but probably even a tighter and richer place.

Weiner: I think there are important differences though. When you talk about comparison of groups you must see what the function was. Just because there was a clustering of people and there were papers coming out does not mean that there were not very important differences. The main task of one institution was building a cyclotron, another's task was educating a generation of students in an unpopular field, another may have been sort of a gathering place for theoreticians from all over the world—these are three quite different functions.

Session III. Plans for Further Research   Charles Weiner, Chairman

Weiner: Shall we get on to the next and final discussion? One question is: given the aim of this conference—to compare overlapping partial perceptions, to offer personal testimony, to explore conflicting interpretations, in order to illuminate the role of theory and theorists as they affected the overall development of nuclear physics as a research field—how far do you think we have come? I ask not because we want to say this was a good day or a bad day, but because we are considering what to do in the future in continuing this kind of activity. And the second area of discussion I would like to open up is: do you have specific suggestions for things that we can profitably do, that either are an extension or continuation of what we are doing now, or perhaps some new departures? I say "we," realizing that we may not be able to carry it out, but it might be some ideal program.

Wheeler: I think this business of identifying our conflicts in point of view was very fruitful. And I can believe that it would be very fruitful to follow this up even more in detail, as for example, to take all the people who refused to accept the compound nucleus model and all the things that they kept looking for as evidence to support their point of view, such as Wigner looking at magnetic moments, and to try to build up the story of how an ideal was kept alive over a long period until at last it could burst into bloom. This, I think, would be very fruitful, and I am sure there are other things of the same kind in these confrontations.

Weisskopf: I still repeat what I said that I think one should not concentrate on these bits of theory because we get a distorted picture. We should have and do have experimentalists here.

Wheeler: The influence of the unexpected, you are saying?

Weisskopf: Yes, actually one can just as well turn it around, and say: what are the important experiments that have influenced theory? This would also be one-sided, but I just repeat my aversion for this guide-line.

Weiner: Our plan for this conference was to have the focus on theory, to bring together primarily theoreticians, but not exclusively (obviously we didn't do that), and then to have another conference, perhaps with emphasis on instrument people. Our first conference, of course, was deliberately a pretty mixed bag, to represent a diversity of points of view. We then felt that it would be good to take pieces of it at a time. To focus one conference on theory, and then as you are saying to have another one to balance it. We might have one on specific experimental developments, or on the instrument designers. It would be a fascinating conference, for example, on the motivations in building large machines, especially the post-war generation of machines, or even on the planning of the large laboratories. Perhaps we need several of these specialized meetings so that we can focus, so that we can get down to cases where people have personal knowledge.

Morrison: Why don't you divide it in a different way? Possibly in order to sub-divide it, make one on a period 1946-1952, but try to cut that period, not just by discipline, but get some theorists, some machine-builders and organizers, and some experimenters, to concentrate on how they saw that period after the war, and what happened until some point at which you want to stop.

Weiner: What do others think about this idea of taking a logical period and bringing together the actors who played various roles and who had various motivations, to talk about the total development in the period?

Caws: I think you would get a much better idea of the influence of theory if you have the experimentalists right there to interact with what the theoreticians are saying. If there is any truth in this division—obviously there is some truth in it—you don't want to exacerbate it by separating the components. And I should think the same would perhaps also go for the instrumentalists.

Weiner: I think the idea of focusing on a time period would eliminate the objection I have that it would get too diffuse. If you take a specific period, it is easier for us to prepare. That is the way historians tend to approach something.

Wheeler: And one could perhaps have the chronology of that period ahead of time, and supplement it by having at least a xerox of the abstract of those papers that are regarded as essential in that period. Then one would have looked at this ahead of time and lived himself back into the period a little more than we have been able to do here.

Weiner: You do feel people would do the work?

Wheeler: If it were just looking at abstracts and not looking at whole papers.

Segrè: To come back to the dividing by periods, I think it is the right thing because I was just thinking in my mind: if you ask the discoverers of the positron, Carl Anderson and Dirac, you will hear two stories that it would be good to hear face to face. They just have nothing in common. I have heard it from Carl Anderson; I haven't heard it from Dirac. But, according to Anderson, Dirac had absolutely no impact whatsoever on his work. He didn't even know it existed.

Weisskopf: There is an interesting corollary earlier: James Franck. The Franck-Hertz experiment was planned and conceived without knowledge of Bohr's paper.

Segrè: Yes, I know that.

Morrison: Just the same, when Anderson got those tracks, he had then been talking to Oppenheimer—I remember hearing that story—and Oppenheimer certainly knew of the positron theory. In fact, he had done a great deal to show that there couldn't be positrons. I think it is true that these people sort of began that way, but it gets caught up into the network. Selection is *a posteriori;* otherwise you would never remember Anderson's experiment if there had not been a theory which is confirmed.

Weiner: The interesting thing in this case is to look at the letters we have from Anderson to Millikan when Millikan was away from Pasadena part of the time. Of course, there is no mention of Dirac or anyone else, and they are strictly talking about cosmic rays in a classic sense.

Segrè: The real complication, according to what Anderson told me, was that the thing was coming from the ground and not from the top. This was really worrying.

Weiner: That is an argument in support of the time period.

Caws: But if I may make one supplementary comment about the time period, I think you are going to have to be prepared to define the time period in a fairly ragged way, that is to say, there may be certain experimental results that have been around for a while, or some suggestive papers that don't get picked up until later. There is a lot of activity for a while in which experimentalists and theoreticians are both at it, but you're going to have to be prepared to go outside the boundaries of that time period with respect to a particular problem, perhaps. And also, presuppositions may get carried into a time period from a previous period and previous problems. In other words, I don't think one ought to be Procrustean about time terms any more than about theoreticians vs experiment.

Weiner: We would set a flexible frame, anyway. From the discussion and from the sorts of things we have covered, my feeling is that a logical period would be the one Phil Morrison suggests, and that is postwar—in your flexible sense—to the early fifties, because this is the transition period we talked about. We want to find out what nuclear structure was doing; we want to talk about this new generation of people. It would involve different people at such a conference, of course, than those of us who are assembled here. So, if there is a feeling that the conference technique is a good one with these modifications, and also the feeling that perhaps the focus of the next conference should be on that period, this is how I interpret . . . .

Weisskopf: Maybe a little longer period than that.

Segrè: I think 1946 to now.

Weisskopf: If you restrict yourself to nuclear structure.

Weiner: We're also interested in the origins of particle physics.

Segrè: There you have to be very careful to get the cosmic ray physicists and the accelerator physicists because this is the transition. For instance, the evolution of Rossi is quite an interesting thing, and the evolution of Carl Anderson—they are particle physicists up to a certain point, and then since they stay with the cosmic rays and they don't go with the machines they are no longer particle physicists because they can't compete. And vice versa, there are the nuclear physicists who become particle physicists because they work with machines. This is quite an interesting thing to see, plus the story of the big national laboratories.

Weisskopf: That's a second subject.

Weiner: A separate subject because it involves administration, funding, politics . . .

Weisskopf: I don't know what principle you have, whether you want continuity, to always go on with nuclear structure, or whether you want to change the subject, and, for example, take the beginnings of particle physics.

Morrison: I feel that the rise of particle physics is an extremely natural topic to get into from the present state, not excluding the problems of nuclear physics as we have seen here, but changing the emphasis and saying: How, when and with whom did particle physics begin to emerge as a separate discipline, presumably sometime in the early fifties.

Weiner: That is precisely what the project is about. We are not writing the history of nuclear structure as such but we are looking at the process of the formation, emergence, development of fields, and the spawning of new fields within those. In that sense, the next look would be at particle physics. I think historians tend to be interested more in these transition points, anyway, so one would get more people working with us and more people interested to help us.

Weisskopf: The origin of particle physics.

Morrison: The rise and decline of particle physics.

Weiner: There is a lot of good stuff on it—the early files for the Rochester Conferences are available, and many other things. The floor is open for methodology, techniques, things we should be doing, things that you would like to do.

Merton: I had one impression of the discussion I'd like to share if we have a few minutes. I might say that the impression was pretty much undermined by remarks made in the last five or ten minutes.

This group evidently has a collective memory: the same events seem to have high visibility and deep significance for all of you. As an outsider who couldn't follow the substantive discussion, I was listening in wholly formalistic terms. From that standpoint, the prime fact was that individuals would pull out the same crucial events from the great mass of events that constituted the development of nuclear physics in this period. Sentences begun by one member of the group would be completed by another. Now, I don't know what comparison with groups in other fields would find. I believe that if there were a conference comparable to this in one of the humanities, you would not find the same developments identified as crucial to the same extent as you have done here.

What was going on here is possible testimony to styles and organizations of interactive and collective thought in one discipline or in a part of it that may not be at all characteristic of what goes on in other fields.

But, as I say, this general impression was undermined by some of your later remarks.

In Unison: Why?

Merton: It was startling to have some of you say toward the end of the discussion: "Well, we must be better prepared for our next meeting. We have forgotten so much of what, in fact, happened. We hadn't done our homework with regard to papers," Yet my notes convey another impression. They indicate that Hans Bethe would declare at one moment that there had been a paper that "appeared, I think, in October 1934" and that Victor Weisskopf would say: "I don't think it was October; it may have been November or December." Now, that's a fairly fine-meshed memory of events that occurred some thirty-five years ago. All apart from dates of publications, you also agreed on what work was being done at the time, again part of a collective memory. The extent and character of this shared memory cannot be taken for granted. They may differ greatly among the disciplines.

Now, how that shared memory came to be and what sustained its texture is itself an interesting question. Note the extent to which you refer to a common literature. I don't know how many papers were published in the period under review in the field of nuclear physics, but I would suppose that it was in the order of ten thousand accumulatively.

In Unison: No, no.

Uhlenbeck: A thousand, at the most.

Weiner: The Beyer bibliography, which was published in 1949, lists not all the papers, but the key papers. I estimate there are 4,000 papers listed there.

Merton: That is of the order of magnitude. My point is that if one goes over the transcript of this conference and notes the allusions to key papers, he will find that you have been dealing with about 100 to 150 papers. This you evidently take as the common core, the shared core. Then, there are all the other papers that may have fit into the texture of the field but the core papers are the common possession of the group.

It might be interesting to have a counterpart group who had worked in nuclear physics, but were not as centrally engaged in it as you, identify the key papers. Would their selections have the same texture, the same nodes? Would they select the same complex of key papers? And, as I say, it would be interesting to have conferences on the history of a set of ideas in other disciplines to see what the character of their collective historical memory is like.

Segrè: I can tell you. If you get the ghosts of the marshals of Napoleon to tell the campaign of Italy, they will just know all the battles and everything that happened.

Zuckerman: One question that comes to mind is whether the experimentalists, looking at the same literature, have picked out the same theoretical items as having been important, and also would the theorists and experimentalists agree on which experiments were also signals.

Weiner: They would, but there are important differences, I think, that we found in interviews. The significance of the neutron to Chadwick was, let us say, a little different than the significance of the neutron to a lot of theorists. In other words, it didn't really solve too many of his problems.

Weisskopf: You [Weiner] have influenced the observation of the speaker [Merton] by your choice of participants here. We are all very close, either collaborating closely, or having been derived from Niels Bohr. I could imagine a choice of participants at the same kind of discussion where you would get somewhat of a different impression, although I believe the first approximations would probably be the same, but there would be more discussion. For example, if Breit and Wigner had come, it would have been livelier.

Weiner: They were invited but could not make it.

Caws: Goldhaber mentioned a change in style since earlier days, in connection with the business of readier speculation now, and I think Professor Bethe said, there are now more physicists. Do the physicists here have any feeling as to the current state of the discipline: whether the same kind of closely-knit network of workers exists, so that 35 years from now you will get the same phenomenon, or was this the golden age of physics that really now is gone? Are things done now in a different way?

Morrison: I believe you would get the same thing if you cut the scope of the investigation far enough. It does throw up these group things and nodes and so on, but their spread is probably somewhat smaller today. I suspect that the theorists there are not as wide. I know they are not as versatile as Hans has been in his time, or Fermi, or even Bohr. But within particle physics you might find people who would be able to give the same kind of characterization.

Wheeler: It would be very interesting to get a Russian that is very active in nuclear physics to comment on the nuclear physics that has come from a very different background.

Weiner: This raises a point. I do think a basic weakness of what we are doing is that we are not really international, except in the origins of the group. But the point is that we refer mostly to the development of nuclear physics in the United States. In the interviews we are trying to balance that somewhat. But perhaps in our definition of the conference about the origin of particle physics, we should say "No holds barred." We would keep the group to a reasonable size, of course, but we would just make sure to invite whomever we feel has something important to contribute.

Weisskopf: In particle physics I think you will get the phenomenon of the unit even stronger.

Bethe: Yes, quite strongly.

Weiner: You mean the idea of the cohesiveness? I'm making a separate point, though, about an international balance.

Wheeler: I wish you could interview Flerov and Smorodinsky after showing them the recordings of what has gone on today ahead of time, to see what they would say about all this.

Segrè: Really, they are pretty isolated systems. They probably found the same things. I know, for instance, that after the war I saw Flerov. Flerov had found spontaneous fission and so on and had made very similar experiments. I didn't know of him; he didn't know of me. I know, for instance, some of the things that Kurtschatov did; I never read them. Kurtschatov never read of what we have done. So it was just two different worlds. You repeated the experiment twice independently.

Bromberg: Dr. Klein and I were wondering whether it would be profitable in a meeting like this to start with an historian's paper, whether a short paper devoted to the period and presenting historical theses would stimulate recollections and ideas any more than they were stimulated today without it. I wonder if people feel that they would remember more and have more ideas with an historical hypothesis.

Weisskopf: You may get more irritated and that way be more talkative.

Weiner: I think people were talkative today. My recommendation would be that any good historical papers on the subject be mailed to all interested people before they come to a conference.

Caws: And you plant them with a few falsehoods just to make sure.

Wheeler: I can believe that this meeting here shows a step-by-step working out of techniques that dig into this history more effectively than it has ever been dug into before, and one can wonder whether it might not be worthwhile to stick to that line for a while further. That is, to find ways, by focusing on this one subject, to analyze it still more in detail, so that then when one started branching out to other fields and other time periods, one would have behind him a well-developed technology of developing the truth. I found today a number of very unexpected things showing up that I had never heard of before. I look forward to the record very much, and I can believe that one, by going over this, at still another level of intensity, could locate still more.

Weisskopf: Instead of going to particle physics?

Wheeler: Right.

Weisskopf: If one chooses your avenue, I think one should very definitely have the experimentalists.

Weiner: These conferences occur at certain nodal points in our research and they sort of stimulate the next stage of the research. My feeling is that we must do more detailed studies as a result of this. And we should have a conference on the origins of particle physics as a prelude to more serious study on that subject. In other words, it doesn't mean that we drop everything now and say that we have done it and deposited a nice little archive. In fact, we should dig in and do a lot of historical studies. And then perhaps in a few years, have another conference to stimulate yet another series of more intensive studies.

Wheeler: So it is good sense to let this cook a while.

Weiner: Right, we have a lot of work to do now.

[END]

APPENDIX I

Prospects for Research on the History of Nuclear Physics Since 1930

Joan Bromberg,
Barry Richman,
Charles Weiner

A discussion paper sent on May 10, 1967 to the participants in advance of the first conference. Copies were also available during the conference and it provided a general framework for the discussion.

I. The Initiation of Nuclear Physics as a Major Research Field

A. Internal Considerations Thirty-five years ago nuclear physics in no way resembled a major research field in physics. Before 1932, according to one of the physicists interviewed, "There were very few people trying to do any nuclear physics." Only four years later, however, the situation had completely changed. By 1936 the young field had developed sufficiently to make Hans Bethe feel the need to summarize what was known in nuclear physics in a review article for the profession. It was, in fact, soon clear that a single article could not do justice to the subject and with the aid of two colleagues at Cornell, Robert Bacher and M. Stanley Livingston, a series of three comprehensive articles was written.[1]

What happened in the early 1930's to bring this field of research to the center of the stage in physics? A number of historic events are worth considering: the discovery of the neutron, the successful operation of particle accelerators at energies high enough to probe the nucleus, the discovery of artificial radioactivity, the Depression, the rise of the Nazis in Germany. But merely to cite these events contributes little to an understanding of the manner in which nuclear physics developed from a cluster of problems to a major research field of high priority.

What is necessary is to determine when, how many, and for what reason physicists were attracted to nuclear studies during the early 1930's. The motivation of the physicists at the University of Rome is illustrative: "At the beginning, the problems tackled by the Rome group were mainly in atomic

1. *Rev. Mod. Phys.*, *8*, 82 (April 1936); *9*, 69 (April 1937); *9*, 245 (July 1937). The "Bethe Bible" as it was and is still affectionately known among physicists was especially significant. One physicist suggests: "It really marked the coming of age, I think, of nuclear physics more than any other single publication that one could easily cite..." Or, as expressed by another: "It came at a time when this field was completely new and in the center of interest..." (Quotations in this study paper which are not attributed by name to a specific individual are excerpted from interviews conducted in preparation for the conference.)

and molecular spectroscopy, but by 1932 it was clear that the future belonged to nuclear physics..."[2] The first major historical problem is simply: What motivated physicists in other fields to enter nuclear physics?

Before Chadwick's discovery of the neutron, construction of a theory of the nucleus was beset with difficulties. Their severity is indicated by the fact that, insofar as the nucleus was concerned, the validity of vital physical principles was doubted. Bohr, for example, called both energy conservation and quantum mechanics into question. "As soon as we inquire more closely into the constitution of even the simplest nuclei, " he wrote, "the present formulation of quantum mechanics fails essentially." Further, "Just as the account of those aspects of atomic constitution essential for the explanation of the ordinary physical and chemical properties of matter implies a renunciation of the classical ideal of causality, the features of atomic stability, still deeper-lying, responsible for the existence and the properties of atomic nuclei, may force us to renounce the very idea of energy balance."[3] It required the discovery of the neutron, together with the postulation of the neutrino and Fermi's theory of beta decay to nullify these possibilities and restore complete faith in the principles challenged. Unlike the transition from sensible to atomic dimensions, and contrary to Bohr's forebodings, the transition from atomic to nuclear dimensions did not demand a new physics.

It is reasonable to assume that physicists are attracted to fields that already boast unexplained phenomena and experimental data, or that somehow show promise for the verification of theoretical calculations. In retrospect it seems that these inducements, concrete results in both experiment and theory, became significant in 1932, the "annus mirabilis" of nuclear physics. In that year Cockcroft and Walton demonstrated that artificially accelerated particles could disintegrate the nuclei of light elements, and Chadwick discovered the neutron. An experimentalist recalls his reaction at the time: "The minute Cockcroft and Walton made their disintegrations, then it began to be quite

2. Emilio Segrè, "The Consequences of the Discovery of the Neutron," *Proc. 10th Intl. Cong. of History of Science,* Ithaca, 1962, 151 (Paris: Hermann, 1964).
3. Niels Bohr, "Chemistry and the Quantum Theory of Atomic Constitution," The Faraday Lecture of 1930, *J. Chem. Soc.,* 379,383, (1932 Part 1).

clear that this was going to be a fruitful game." The discovery of the neutron meant that no longer was it necessary to assume that the electron was a constituent of the nucleus. Therefore the powerful theoretical methods of the new quantum mechanics could be applied to the nucleus. A theoretician says, "I went into nuclear physics only after 1932...after the discovery of the neutron in 1932, it was in a general way clear what had to be done...I cannot invent something out of nothing..."In the same vein, another theoretician recalls, "For me [nuclear physics] started with Heisenberg's paper [published in July, 1932] ... [he] pointed out that now that the neutron has been discovered, one can think of starting a theory of the nucleus. This impressed me very much." Another man, who received his doctorate in 1933, explains his interest at the time: "I believe that the factor was the idea that nuclear physics was the exciting forefront in physics, the exciting area for making progress in physics, and that out of the interactions between particles as shown by scattering and nuclear binding one could hope to learn the forces between elementary particles."

These comments indicate some of the major motivations of those who came into nuclear physics after 1932. There is also good evidence that the small group of physicists engaged in research on the nucleus during the preceding five years were hopeful that they *could* develop either experimental or theoretical techniques that would satisfy their desire to understand nuclear forces or to apply quantum mechanics to the nucleus.

These initial impressions suggest answers to a number of questions and raise still others: Were any theoreticians attracted to nuclear physics at the time when it was thought possible that entirely new physical laws might be involved? Who and for what reasons? What motivated those physicists who entered the field after the discovery of the neutron?

What attracted scientists to experimental nuclear physics? How were individual entries distributed in time? Did the novices address themselves to problems quite different from those with which they were familiar? Or did they pursue already chosen "themes," preoccupations which they felt could be more fruitfully attacked in this new area?

One eminent physicist whose interest in the field was particularly ephemeral is Werner Heisenberg. When respected leaders, such as Heisenberg, left the field because it did not continue to satisfy their research interests, was there a related exodus of respectful followers? It would also be worth knowing whether it proved easier for theoreticians than experimentalists to move into and out of nuclear physics.

All of these questions can be approached through oral history interviews and by studies of what these physicists wrote and talked about in the journals, at conferences, and in personal correspondence.

Despite some defections, nuclear physics kept growing. By late 1937, many structural features of the nucleus were discerned. Its composition of neutrons and protons was accepted. Some indications of shell structure had been found, while the slow-neutron experiments of Fermi and others had given rise to the model of the compound nucleus. The proton-proton force had been discovered, and the concept of charge independence proposed.

At the same time, a series of important discoveries and innovations opened the way to a rich harvest of information. Among these was the discovery of deuterium (1932), artificial radioactivity (1934), the improvement of detection techniques, and artificial acceleration of particles to useful and steadily increasing energies by a variety of methods. The field continued to be fruitful and its prestige continued to increase within the physics community.

B. The Social Aspects An important element in the maturation of any field is its internal system of communications. How were the results of current research diffused in nuclear physics? What was the relative significance of the journal literature, international conferences, national and regional meetings, summer schools, journal clubs, colloquia, exchanges of personnel among research centers, and personal correspondence? The interviews offer some insight here, especially by suggesting documentary source materials for studies in this category. There are indications that regular "journal club" meetings were held at almost all the major research centers in the United States. In England the Cavendish Laboratory's version of a journal club was the Kapitza Club, the minutes of which have been preserved. All the documents of such institutions must be examined.

Other major factors in the development of a field are the availability, scale, and cost of apparatus, instrumentation and materials. Nuclear physics required radioactive sources, equipment to accelerate particles, and radiation detectors. The technological complexity, scale, and expense of experimental research increased steadily. Where did the funds come from? What arguments were used to obtain them? What were the motivations of the donors? What effect did the Depression have on the source of funds? The full story must be sought in the records of government agencies, private foundations, scientific organizations, and physics departments, but interviews and personal correspondence from the period have already established some interesting and even unlikely sources of support. For example, medical foundations supported nuclear physics research at the California Institute of Technology and at Berkeley because of the possibility of using the apparatus for treatment of cancer victims.

The commercial availability of components, or the ability to make them, understandably influenced the design of apparatus and ultimately the choice of research problems. A distinct theme expressed by many of the men interviewed is that the high level of technology in the United States was a boon to experimentalists, who could buy vacuum tubes at the corner drug store and condensers at radio repair shops. Further, many of the men interviewed (experimentalists *and* theoreticians) had "tinkered" with radios as boys and had entered college with engineering as their original career goal. This high technological level and consciousness thus appears to have been significant whether it was manifested in the ready availability of cheap replacement parts, the construction of high-voltage sources, or the awareness of the capabilities of contemporary technology.

How important was this technological background to the overall development of nuclear physics? Was the situation different in Europe? What, in short, were the differences in contemporaneous technologies; what were the consequences for various centers of nuclear physics research?

Nuclear physicists who experienced it need not be reminded of the effect that the Depression had on their careers. Jobs were scarce for everyone in physics, and nuclear physics was not generally regarded as a particularly valuable specialty for government or industrial employment. The university was the natural home for nuclear physicists, with the outstanding exception of the Department of Terrestrial Magnetism of the Carnegie Institution of Washington. The general reduction in university budgets and junior faculty most affected the younger men who had recently earned their doctorates in physics. Their problem was not where to obtain an opportunity in *nuclear* physics, but rather where to get any employment whatsoever in physics.

Some of those interviewed recalled that many of their classmates left physics after graduate school and that others found employment on seismic prospecting gangs in the oil fields of the southwest. There was a tendency for students to select graduate schools on the basis of the availability and amount of financial assistance offered, rather than on the reputation or specialties of the faculty. And once in graduate school, there was no hurry to leave.

As yet far too little is known about the effects of the Depression on science in general in the United States and Europe, much less on physics fields and specialties. Since nuclear physics had a high intellectual priority within the physics community by the mid-1930's, did a physicist stand a better chance of obtaining a fellowship, assistantship, or faculty position if he was involved in nuclear physics? Physics department and foundation records may yield the answer.

Still another factor in the field's development was the unprecedented mobility of physicists and potential physicists caused by the policies of the Nazi regime. Deprived of university positions because of their failure to meet genealogical or political criteria, increasing numbers of scholars fled the Fascist countries and made their way to other lands. Yet, the diffusion of highly trained physicists occurred at the time when nuclear physics was taking shape as a major research field. It is problematic whether or not the field would have reaped greater benefit from political stability. What impact or influence did the emigrés have on nuclear physics in the countries to which they emigrated? How were they received? How were they influenced by their new environment?

The interviews include descriptions of the circumstances under which nuclear physicists left Germany and Italy, and their expectations and experiences as guests and permanent settlers in Copenhagen, London, Cambridge, Manchester, Ithaca, Rochester, Urbana, Princeton and Berkeley. A study based on documentary source materials has been initiated to assess the impact of the emigrés on the development of nuclear physics in England and the United States. Did the emigrés import whole a tradition of theoretical physics that did not exist in the United States? Or had American theoretical and experimental physics already come of age by the early 1930's? Were there national differences in the relationship between theory and experiment? In attempting to answer it is necessary to distinguish between national style and personal style and to consider the problems encountered in a specific field of physics during a specific period. Some lines of inquiry related to the connections between theory, experiment and technique in nuclear physics are introduced in the next section.

II. The Interaction of Theory and Experiment

Approached from almost any perspective of interest, an examination of the interplay of theory and experiment in nuclear physics must prove rewarding. The field affords an exceptional opportunity to follow the former as it becomes increasingly comprehensive, the latter as it grows progressively more precise. Consider this perspective: What theoretical considerations affected, first, the design of particle accelerators and, second, the design of experiments as the science developed?

This is a question suggested by a view expressed in the interviews, namely, that prewar theories were *not* influential as guides to experiment. An experimentalist was asked, "How did the current theoretical concepts of the nucleus, the models of the nucleus that existed at that time, affect the work that was done in the period?" He answered, "One knew that the nucleus was made out of neutrons and protons, but one didn't take anything else seriously at all..." [Question] "Did the compound-nucleus model affect the interpretation of experimental work?" [Answer] "Not much, because it was not something on which you really could put your hands and your teeth very clearly..." [Question] "So what was the picture...you had some picture, I assume, of the nucleus in your mind?. [Answer] "We didn't have much of a picture, no."

How, then was laboratory work guided? The simplest assumption is that the goal in *machine design* in this prewar period was just the production of artificially accelerated particles with sufficient energies to probe the nucleus, while the goal in *experiment design* was largely amassing information. In Rome the systematic slow-neutron bombardment of each entry in the periodic table is exemplary, and similar work was done with the Berkeley cyclotron in the 1930's. In contrast, the relation between theory and experiment (and accelerator building) seems qualitatively different in the early fifties.

Compare the following statements in which two physicists, recently interviewed for our project, explain their reasons for designing accelerators. The first comments on his motivation for designing an accelerator in 1926:

I said "Oh, I want to go and do high voltages with Rutherford. Somebody has got to make high speed protons, artificial alpha particles. We know that if you put a million volts on a charger, you'd get something that's more or less equivalent to probe the nucleus. Protons and electrons somehow are stuck together, and I want to do that."

The second physicist explains his motivation for building a new accelerator thirty years later:

We had by this time [1956] ...built three machines...And although we were very happy with these, we were continually being frustrated. You see, by this time we were beginning to understand a little of the theories. By this time we were beginning to see what kinds of questions [the theoreticians] would like to have answered. "What are they talking about? What measurements can one make that bear on the theory?"

Do these two statements adequately represent the end points of this 30-year period? What of the progress in between? What part, for example, was played simply by lack of information about nuclear forces?

Pursuing the relation of experiment and theory, what qualities of theory are useful to experimenters? Certainly the ability to predict, to get some indication that an effect is worth measuring. But there are other possibilities. One is the perhaps intangible benefit of the theory that aids intuition. An experimentalist's wry view:

> From the point of view of making a poor, uneducated experimentalist understand what those guys are saying...[the collective model is] a godsend, just a beautiful little model almost as good as the billiard ball. You can make calculations with it without understanding anything...You make experiments, you make predictions, all by yourself.

As well as illuminating the role of models in physics, this statement introduces another group of germane considerations. There is, in general, a division of labor in physics between theorists and experimentalists. Who has borne the responsibility for choosing the theoretical questions to be tested in nuclear experiments, the chief experimentalists or the theorists with whom they were in contact?

The physicist quoted suggests elsewhere in his interview that there has been a tendency for nuclear theory to become more accessible intuitively. He sees this as the convergence of two trends: on the one hand, intuition has become educated to modern concepts; on the other, the theories have become more physical. If this suggestion is supportable, it may be found that experimentalists are becoming increasingly able to use their own ideas as bases for experiments, and that, in fact, they are increasingly doing so.

Another experimentalist, however, holds a diametrically opposed view. Experimentalists, he suggests, are increasingly working on the hypotheses of others. The factors to which he attributes this are social ones: greater facility in communication and a shift in the criteria by which the community judges a scientist's stature.

> Nowadays the ideas seem to mean less because they are sort of in the air. People work on each other's ideas...The modern experimenter finds it safer to work on a theoretician's idea. This is a real switch compared with my young days, and this, to me, is in a way a very dangerous switch for physics... you have to be competent and accurate and fast. These are still necessary qualities. But you don't have to have ideas.

Are there in fact any general patterns of change in the way ideas for experiments are generated and, if there are, what determine them?

The other side of the effect of theory on experiment is just how the increasing store of experimental data affected the acceptance or replacement of theories. Here the historian will have a rich source of information in the history of models. Representative questions abound. How decisive was the experimental evidence for the ascendency of the compound nucleus model over the early shell model? As is well known, the shell model was re-introduced after the war. With the *ad hoc* assumption of the spin-orbit force, it displayed impressive agreement with experiment. Yet there is evidence of a variety of negative reactions to revised shell theory. Where resistance was encountered, whenever it was encountered, what were its sources? Here it is necessary to consider the assessments of scientists of the relative *importance* of experimental confirmation of a theory, on the one hand, and theoretical or experimental *justification* on the other.

One of the most striking aspects of nuclear physics is its introduction of instruments of unprecedented expense and size, and one of the most obvious questions for the historian is the effect of this new kind of instrumentation on the character of science.

Does the new scale of operations encourage or intensify a tendency to "play out" a piece of equipment, to do every experiment that can profitably be done with it, even at the expense of the physics involved? The physicists questioned on this point were by no means unanimously agreed that this is the case. One experimentalist pointed out that the same tendency had been characteristic, for example, of the diffraction grating.

While the relation of technique to experiment remains problematic, its effect in enforcing a new organization of work is unmistakable. One of the most interesting developments is a new alienation of the physicist from his tools. "You have technicians who run alpha-particle spectrometers and beta-ray spectrometers and gamma-ray coincidence circuits. And the people working in the field...are putting things into computers and analyzing the print-out and they are pretty well disconnected from the experimental side of it..." And he added, "If I were a graduate student coming into physics now, I wouldn't work in my own group because I would find it too dull." Here is displayed the clear (if not unanimously held) view that the time and the virtues of the time when one ran 100 yards to measure an activity, analyzed data with a slide rule, and learned the art of improvisation are gone. Things have changed since the '30's.

III. Nuclear Physics as a Research Field in the Postwar Period

What was the effect of World War II on nuclear physics? Replies to this question were, in several respects, surprisingly uniform. As a result of the war:
(a) Physicists were not afraid to ask expensive questions.
(b) They had acquired confidence in their ability to tackle large-scale difficult projects.
(c) Many learned sophisticated electronic techniques in radar that were applicable to nuclear physics.

As for the importance to nuclear physics of nuclear fission developments *per se*, there was less general agreement. Most felt that the work on fission permitted a good, if limited, test of the predictive ability of nuclear models, increased knowledge of nuclear reactions, and provided new isotopes and new sources of neutrons for experimental research. However, the relative importance of these developments as contributions to nuclear physics was minimized by several physicists.

Some men felt that the war was a hiatus in the development of nuclear physics, although one of them felt that the delay might have been beneficial because it forced physicists to get away from the subject, introducing them to men and techniques they might otherwise never have known. At the M.I.T. radar laboratory and later at Los Alamos, national, regional, and age barriers were overcome and a large group of physicists and technicians shared the new experience of large-scale team research and development. Yet few of the men who called themselves nuclear physicists either before or after the war felt that they were doing nuclear physics during the war at Los Alamos.

Another point that needs clarification is the basis of the postwar consensus that research funds for nuclear physics would be plentiful. How did the outlook for financial support influence selection of research problems? What other factors influenced the selection of problems?

The interviews offer some partial answers:

For the first time a lot of money and a lot of people were concentrated and this led to very rapid technical developments. Plus the developments in radar microwaves, electronics generally. Plus the availability thereafter of large amounts of money; the reputation of the projects made research money easy to get. [Question] Would you characterize this [war period] as a "tooling-up" period? [Answer] Mostly. A lot of detail was filled in in nuclear physics and not very much was learned in principle.

Yet, as the war was ending, another man recalls:

It didn't occur to me that we would get the huge support for physics that later turned out to be true. I thought we would go back from the wartime era when we had unlimited amounts of money to the system we had before the war when we didn't get money at all.

But certainly new concepts of organization for research, new sources of funds, new principles of design, and new technology were not the only factors that motivated physicists to build new large scale apparatus. To what use would these machines be put? The answer of one physicist represents a motivation that was apparently shared by many of his colleagues:

By that time it was clearly understood there were limits as to what you could find out [about nuclear forces] with the energies of protons and neutrons that were available previously. And the thing to do was to build high energy machines to get high energy protons and neutrons. You were attacking old problems still.

Were the new high energy accelerators designed to tackle long standing problems relating to the nature of nuclear forces? Was there a special sense of urgency to reach high energies because of an intensification of theoreticians' interest in nuclear forces? Did experimentalists share these motivations?

In considering the development of nuclear physics as a research field after World War II, attention must be given to many factors, including: the new patterns and scale of government support of research; the emergence of national and international laboratories devoted to nuclear physics, the participation of industry in developing crucial detection techniques. Moreover any of several internal developments—the successes of postwar structural models of the nucleus, increased understanding of nuclear forces, the power of many-body theory—could structure several full days of discussion.

One class of questions, however, seems of special interest at this stage in our efforts to chart the course of nuclear physics as a field within physics. They concern the development in the late 1940's of subnuclear or elementary particle physics as a field distinct from nuclear physics. It is a field that emerged as nuclear physics became mature and we are thus particularly concerned here with the relationship of this new field to nuclear physics.

Elementary particle (high energy) physics is clearly a natural extension of nuclear physics in the sense that higher energy accelerators and more sophisticated detection devices made it possible. But if that is all, why has it developed as a separate field? Again the motivations of those entering the field are vital data, just as they were in examining the inception of nuclear physics. For example, why, instead of coming into nuclear physics, were young theoreticians in the postwar period attracted in the late 1940's to electrodynamics, after that to field theory and finally to elementary particles? Why did many experimentalists and theoreticians (almost all of the men interviewed thus far) leave nuclear physics, and what effect did this have? Why did those who remained in nuclear physics do so? Why did new people enter the field in the past twenty years? What remains in nuclear physics to do?

This conference postulates the suitability of nuclear physics as a legitimate, well-defined object of historical inquiry. The questions raised in the preceding sections sketch points for such inquiry up to the departure of particle physics. It is to be expected that much that is significant has been touched on only superficially, some important issues have been left out, and some matters of minor consequence treated to excess. Conference participants are earnestly requested to express their views of the balance and to recommend departures they feel will enrich the joint project.

APPENDIX II
Data Resources for Research on the History of Nuclear Physics

The following are the major data files developed during the research project related to the conferences. Detailed descriptions, the files themselves, and a series of statistical tabulations based on many of them are available in the offices of the Center for History of Physics at the American Institute of Physics, 335 East 45 Street, New York, N. Y. 10017. Examples of the use of some of the data are shown in the table and figure on page 268 which depict the rapid growth of nuclear physics in the United States during the 1930s. The data files based on publications in *The Physical Review* (described in items 1-3 below) were compiled by Henry Small, in consultation with Beverly Porter, and with the assistance of Dorothy Schreibersdorf.

1. Nuclear Physics Bibliographic File: *The Physical Review,* 1927-1940.

— 5 x 8 cards; includes all nuclear physics papers, letters and abstracts published in *The Physical Review* during the period. Data available on the cards include the following:
   a. Year, volume and page numbers of publication
   b. Title of publication
   c. Authors
   d. Institutions and departments of authors
   e. Dates of submission and publication
   f. Type of publication: paper, letter, abstract
   g. Acknowledgments
   h. Funding.

2. Listing of all publications in *The Physical Review,* 1927-1940.

— includes separate listings for papers, letters and abstracts. No titles, but *The Physical Review* page number, author and institution of each publication are indicated.

3. Other Files derived partially or wholly from the above.

   a. Nuclear Physics Population Listing

   — alphabetic listing of all individuals publishing in nuclear physics in *The Physical Review,* 1927-1940. Data includes:

1. Institutions from which individual published by year
2. Number of publications individual produced by year
3. Year of birth, year, university and field of highest degree. If highest degree was in physics post-1927, the sub-field of physics is also indicated.

b. Nuclear Physics Institutional Roster

— annual roster, 1927-1940, of institutions involved in nuclear physics with listings of authors who published from each institution each year.

c. Nuclear Physics Funding Listing

— annual listing of funded nuclear physics publications, including funding agency, author, institution and subject of publication.

d. Nuclear Physics Acknowledgment Listing

— annual listing of nuclear physics publications acknowledging some type of assistance, including author and his institution, person or institution acknowledged, and reason.

4. Nuclear Physics PhD Files. Compiled by Beverly Porter with the assistance of Dorothy Schreibersdorf.

a. U. S. Nuclear Physics PhDs, 1927-1942.

— on cards, ordered by institution;
— includes name, date and title of dissertation and institution from which degree was received. (Original source was M. Lois Marckworth, *Dissertations in Physics: An Indexed Bibliography of All Doctoral Theses Accepted by American Universities, 1861-1959.* Stanford: Stanford University Press, 1961.) A chronological listing of the same material is also available.

b. Listing of U. S. Nuclear Physics PhDs who published in nuclear physics in *The Physical Review,* 1927-1940.

— includes date and institution of degree and dates and institutions of publication.

c. Nuclear Physics PhDs who did not publish in nuclear physics in *The Physical Review*, 1927-1940, with information on their activities where available.

5. Chronological listing, by physics sub-field, of all U.S. physics PhDs, 1928-1942. Compiled by Elizabeth Garber.

— includes institution and title of dissertation (also derived from Marckworth, see above).

6. National Research Fellows in Physics, 1919-1942.

— on cards;
— includes university, date and subject of degree, title of fellowship research and years and places of fellowship. Those working on nuclear physics topics have been maintained separately.

7. Rockefeller Foundation Fellows in Physics, 1917-1950.

— on cards;
— includes university, date and subject of degree, years and place of fellowship.

8. Emigre Physicists in the 1930s.

— on cards;
— includes biographical, educational and career data;
— compiled from records of Academic Assistance Council (England), Emergency Committee in Aid of Displaced Foreign Scholars (U.S.), and other sources.

9. Bibliography of Published Papers on Nuclear Structure, 1911-1940. Compiled from major abstracting journals by Henry Small.

10. Bibliography and Abstracts of Articles on Physics in *The New York Times,* 1930-1940, and in the *Readers' Guide to Periodical Literature,* 1929-1941.

– on cards.

11. Chronological Listing of Nuclear Physics Discoveries, 1930-1950.

– on cards;
– includes full literature reference. Contains many gaps. Compiled by John Titus.

12. Particle Accelerators, late 1920s-1940s.

– on cards;
– listed by institution, type of machine, energy, particles accelerated, date of construction. Literature references. Not complete. Compiled by Elizabeth Garber.

GROWTH OF NUCLEAR PHYSICS IN THE UNITED STATES

Nuclear Physics Publications, Authors and Institutions Appearing in The Physical Review, 1927-1940

Year	Nuclear Physics Publications	Nuclear Physics Authors	Nuclear Physics Institutions
1927	13	15	10
1928	19	20	9
1929	23	23	15
1930	48	48	23
1931	72	65	26
1932	77	73	26
1933	156	114	27
1934	194	158	33
1935	206	170	42
1936	268	200	41
1937	301	234	55
1938	283	246	55
1939	371	320	73
1940	359	358	85

Percentage Nuclear Physics Publications in *The Physical Review,* 1927-1940

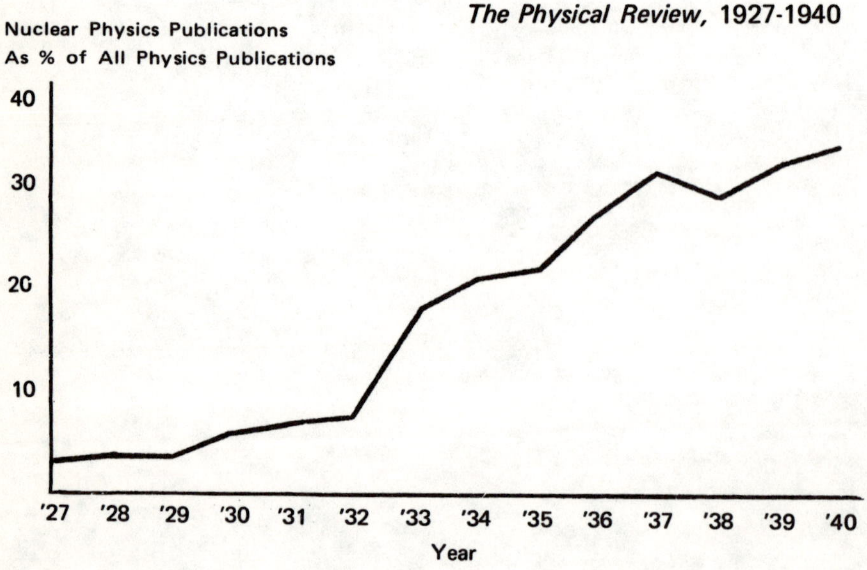

APPENDIX III
Original Source Materials for the History of Nuclear Physics

A. Documentary Materials

One of the major aims of the conferences and related research project was to identify, locate, and assist in the preservation of documentary source materials which could provide the basis for scholarly studies of the development of nuclear physics. The interviews and the discussions at the conferences provided valuable information on materials in private hands, in the files of institutions, or in archives. A systematic effort was and is being made to examine such materials and to insure that they are properly preserved in appropriate repositories, cataloged, and made available for research. Many of these letters, diaries, notebooks, manuscripts, and institutional records in turn were valuable sources for the research preparation for subsequent interviews and for the second conference. Information on the location, nature and scope of these materials is recorded in the National Catalog of Sources for History of Physics maintained by the AIP Center for History of Physics, and is available by writing to the Center at 335 East 45 Street, New York, N. Y. 10017.

Among the collections relevant to nuclear physics which are described in the Catalog are the personal papers of the following scientists:

Samuel K. Allison	William A. Fowler	George Pegram
Jesse W. Beams	George Gamow	Ernest Rutherford
Hans A. Bethe	Samuel Goudsmit	Otto Stern
Niels Bohr	Lew Kowarski	Leo Szilard
John Cockcroft	Ernest Lawrence	Merle Tuve
Robley Evans	Gilbert N. Lewis	Eugene Wigner
Enrico Fermi	Lise Meitner	C.T.R. Wilson
	Robert Oppenheimer	

The Catalog also contains information on the archives of institutions including materials on the physics departments at several U.S. universities, such as California, California Institute of Technology, Chicago, Columbia, Illinois, Michigan, Minnesota, Princeton, and Wisconsin.

More detailed information on the contents and availability for research of these collections and others related to the history of nuclear physics is available from the Center. (See also, *A Selection of Manuscript Collections at American Repositories,* prepared by Joan N. Warnow. National Catalog of Sources for History of Physics. Report No. 1. AIP Pub. R-217. Several relevant collections are listed in *Sources for History of Quantum Physics: An Inventory and Report,* Thomas S. Kuhn, John L. Heilbron, Paul Forman and Lini Allen, eds., Philadelphia: The American Philosophical Society, 1967.)

B. Oral History Interviews

Nearly 200 hours of tape-recorded interviews with 41 physicists from 21 institutions in the United States and in Europe have been conducted in connection with the conferences and research project. Preparation for the interviews included study of available published and unpublished documents on the life, work and writings of the individual, the institutional environments in which he worked, and the intersection of his career with important developments in the history of nuclear physics. In most cases the interviews were conducted in the homes or offices of the interviewees where access to supplementary written documentation was available. In almost all cases the autobiographical approach was used, with coverage of family background, early life, schooling, development of interest in science, subsequent education, career decisions, choice of research problems, circumstances surrounding work on specific problems, awareness and reaction to other work in the field.

According to the procedures of the oral history program of the AIP Center for History of Physics the tapes are transcribed. Then, after the transcripts are checked by the interviewer, they are sent to the interviewee who is requested to review them for accuracy, to make additions if he desires, and to provide clarification of matters that may have come to light through the study of source materials since the interview. The retyped, final transcript is then deposited in the Center's Niels Bohr Library along with related bibliographies and documentary source materials, are made available for research by qualified scholars, subject to any restrictions placed by the interviewee. Elspeth Hart has been responsible for the detailed work involved in this process of transcribing recorded tape into useful research material.

The interviews provide a rich, diversified, personal record of the human and social aspects of nuclear physics, and contain information on the social and intellectual development of the field which is an important supplement to written documentary materials. Interviews with the individuals listed here are not all completed. Some are still in various stages of processing, and in some cases further sessions are planned. Additional interviews with other physicists will be conducted in the future. Information on the availability of transcripts for research use and on other relevant interviews may be obtained from the AIP Center for History of Physics.

The following individuals have been interviewed in connection with the history of nuclear physics conferences and research project:

Luis W. Alvarez
Edoardo Amaldi
Carl D. Anderson
Robert F. Bacher
Guido Beck
Hans A. Bethe
Felix Bloch
James Chadwick
John D. Cockcroft
Edward U. Condon
Philip I. Dee
Lee A. DuBridge
Robley D. Evans
Norman Feather
William A. Fowler
Otto R. Frisch
George Gamow
Wolfgang Gentner
Maurice Goldhaber
Lew Kowarski
L. Jackson Laslett
Charles C. Lauritsen
Thomas Lauritsen
Hilde Levi
M. Stanley Livingston
Edwin M. McMillan
Christian Møller
Philip A. Morrison
Seth N. Neddermeyer
Giuseppe Occhialini
Mark Oliphant
Rudolf E. Peierls
Léon Rosenfeld
Emilio G. Segrè
Robert Serber
Merle A. Tuve
Sten von Friesen
Victor F. Weisskopf
John A. Wheeler
Milton G. White
Eugene P. Wigner

AIP Conference Proceedings

		Published	Price
No. 1	Feedback and Dynamic Control of Plasmas (Princeton 1970)	1970	$11.00
No. 2	Particles and Fields - 1971 (Rochester 1971)	1971	$10.00
No. 3	Thermal Expansion - 1971 (Corning 1971)	1972	$10.00
No. 4	Superconductivity in d- and f-Band Metals (Rochester 1971)	1972	$11.00
No. 5	Magnetism and Magnetic Materials - 1971 (2 parts) (Chicago 1971)	1972	$17.50
No. 6	Particle Physics (Irvine 1971)	1972	$ 8.25
No. 7	Exploring the History of Nuclear Physics (Brookline 1967 and 1969)	1972	$10.00
No. 8	Experimental Meson Spectroscopy - 1972 (Philadelphia 1972)	1972	$11.00
No. 9	Cyclotrons - 1972 (Vancouver 1972)	1972	
No. 10	Magnetism and Magnetic Materials - 1972	1973	$17.50